Technology in a Globalizing World

www.novapublishers.com

Technology in a Globalizing World

Facial Recognition Technology: Usage by Federal Law Enforcement
Mari F Burke (Editor)
2022. ISBN: 979-8-88697-124-8 (Hardcover)
2022. ISBN: 979-8-88697-175-0 (eBook)

Nonlinear Systems: Chaos, Advanced Control and Application Perspectives
Piyush Pratap Singh, PhD
2022. ISBN: 978-1-68507-660-3 (Softcover)
2022. ISBN: 979-8-88697-001-2 (eBook)

Multidisciplinary Science and Advanced Technologies
Fernando Gomes, PhD (Editor)
Kaushik Pal, PhD (Editor)
Thinakaran Narayanan (Editor)
2021. ISBN: 978-1-53618-959-9 (Hardcover)
2021. ISBN: 978-1-53619-198-1 (eBook)

Issues with Facial Recognition Technology
Warren Lambert (Editor)
2020. ISBN: 978-1-53618-973-5 (Hardcover)
2020. ISBN: 978-1-53619-000-7 (eBook)

The Pulp and Paper Industry: Production, Management and Technology
Mette P. Kristensen (Editor)
2020. ISBN: 978-1-53617-919-4 (eBook)

More information about this series can be found at
https://novapublishers.com/product-category/series/technology-in-a-globalizing-world/

Kamran Kheiralipour

Sustainable Production

Definitions, Aspects, and Elements

Copyright © 2022 by Nova Science Publishers, Inc.
DOI: https://doi.org/10.52305/PMEU7193

All rights reserved. No part of this book may be reproduced, stored in a retrieval system or transmitted in any form or by any means: electronic, electrostatic, magnetic, tape, mechanical photocopying, recording or otherwise without the written permission of the Publisher.

We have partnered with Copyright Clearance Center to make it easy for you to obtain permissions to reuse content from this publication. Simply navigate to this publication's page on Nova's website and locate the "Get Permission" button below the title description. This button is linked directly to the title's permission page on copyright.com. Alternatively, you can visit copyright.com and search by title, ISBN, or ISSN.

For further questions about using the service on copyright.com, please contact:
Copyright Clearance Center
Phone: +1-(978) 750-8400   Fax: +1-(978) 750-4470   E-mail: info@copyright.com

NOTICE TO THE READER

The Publisher has taken reasonable care in the preparation of this book, but makes no expressed or implied warranty of any kind and assumes no responsibility for any errors or omissions. No liability is assumed for incidental or consequential damages in connection with or arising out of information contained in this book. The Publisher shall not be liable for any special, consequential, or exemplary damages resulting, in whole or in part, from the readers' use of, or reliance upon, this material. Any parts of this book based on government reports are so indicated and copyright is claimed for those parts to the extent applicable to compilations of such works.

Independent verification should be sought for any data, advice or recommendations contained in this book. In addition, no responsibility is assumed by the Publisher for any injury and/or damage to persons or property arising from any methods, products, instructions, ideas or otherwise contained in this publication.

This publication is designed to provide accurate and authoritative information with regard to the subject matter covered herein. It is sold with the clear understanding that the Publisher is not engaged in rendering legal or any other professional services. If legal or any other expert assistance is required, the services of a competent person should be sought. FROM A DECLARATION OF PARTICIPANTS JOINTLY ADOPTED BY A COMMITTEE OF THE AMERICAN BAR ASSOCIATION AND A COMMITTEE OF PUBLISHERS.

Additional color graphics may be available in the e-book version of this book.

Library of Congress Cataloging-in-Publication Data

ISBN: 979-8-88697-057-9

Published by Nova Science Publishers, Inc. † New York

Contents

Preface		vii
Chapter 1	**Production**	1
Chapter 2	**Design**	11
Chapter 3	**Implementation**	21
Chapter 4	**Management**	29
Chapter 5	**Sustainability**	41
Chapter 6	**Economic Aspect**	49
Chapter 7	**Environmental Aspect**	55
Chapter 8	**Social Aspect**	63
Chapter 9	**Technical Aspect**	69
References		85
About the Author		115
Index		117

Preface

In sustainable production, the first goal of each production activity must be to create an overall positive effect on the world so environmental, social, and economic aspects must be respected simultaneously as well as the technical aspects.

Production is the act to produce a product in form of a commodity or service. Production has high importance in each country from economic, technical, social, and environmental aspects. On another side, sustainability is a hyper-disciplinary issue which became one of the main concerns across the world. It encompasses sections of society as academic fields, and all companies, factories, organization systems, and consumers. It is a hyper-disciplinary issue because it involves all aspects of human life as economic, environmental, and social aspects. Therefore, it is necessary to define and explain the various considerations and elements of sustainable production. The goal of the book is to define all aspects and elements of sustainable production for readers. Sustainable production has different aspects as technical, economic, environmental, and social. Each of these aspects has different elements in sustainable production. In this book, different elements in each aspect have been described and many of them are affected by each other. For the production of a product, all or some of the elements may be considered to be respected. This book can assist the readers in evaluating the sustainability elements in their lives, jobs, and researches. The readers may be students, researchers, manufacturers, and those who conduct any producing or servicing activity.

Sustainability is one of the main important concerns in the world. Like human prosperity, sustainability has no end that is, the more prosperity, the better. Sustainability is a path that human beings must move on its road to improve the state of sustainability.

Sustainability is a hyper-disciplinary issue because it has different aspects, relates to all aspects of human life, and all studied fields can help to be closed to sustainability. The first seen view in human life is the economy because it determines the purchasing power of the people. But, the economic view of human life is looking for economical profits via cost-benefit analysis and doesn't emphasize preserving the environment. Sustainability wants to keep permanent continuing progress for all aspects of human life, so it includes a sustainable environment, sustainable economic, and sustainable social aspects. It means that the universe must be conserved in point of the three aspects. Sustainability is achievable by helping with technical aspects of human activities besides them.

Production activities are those to produce products, commodities, or services. So, a great part of human actions in each country belongs to production activities. Another issue in production is its vast dimensions. Therefore, it has high importance and it is necessary to define and explain the various elements affecting production.

Due to the high importance of production and sustainability issues, this book has been written to discuss the different aspects and elements of the ongoing sustainable production path. The purpose of the book is to explain the technical, environmental, social, and economic aspects and elements that all help to reach sustainability.

This book involves different elements affecting sustainable production. The book wants to provide introductory information in brief, for producers, who start a production activity, the people interested in green production, and consumers of green products for ongoing sustainability road. The provided information in this book stimulates the reader to study more books and articles about the effective elements to reach sustainable production which their introductions can be found in this book. By completing the information and knowledge in this road, the readers become great advocates of sustainable production as well as sustainability.

The book includes 10 chapters to provide information in the field of production, sustainability, and different aspects and elements of sustainable production.

Chapter 2 wants to create an insight bout production and its aspects and importance for readers. The definition of production has been provided in Chapter 2. In this chapter, the goal of production and its importance have been described. Different production factors, phases, and aspects have been explained.

For the production of an especial product, either commodity or service, firstly a proper design must be prepared for the product. The designing process is very important because it is the first step in the production of a product. Different factors must be respected to provide an acceptable design. Chapter

3 explains the design process, design requirements, design importance, and design types.

After designing the design, it must be correctly implemented, including manufacturing, construction, and so on. Chapter 4 describes the implementation process and its requirements.

To reach the best designs and to conduct the implementation of the process in the production of a product, the process must be properly managed. In Chapter 5 the definition of management and its requirements and types have been provided.

After providing the information about designing, implementing, and managing to produce a product, the sustainability issue has been described in Chapter 6. Sustainable production was explained at the end of this chapter.

As mentioned, different economic, environmental, technical, and social aspects must assist sustainable production to reach sustainability. Each of these aspects has different elements in sustainable production. Many of the elements are affected by each other, for example, there are some technical element effects of some of the economic and environmental elements. Based on the product types, nature, and complexity all or many of the elements are required in producing process. Sustainable production elements including economic, environmental, technical, and social aspects and elements have been presented in Chapters 7-10.

The correct definitions for mentioned keywords in the book such as sustainability, design, implementation, management, production, service, economy, environment, social, and so on, have been presented. For each keyword, some references have been provided for more study about them.

Hope the book provides beneficial information about sustainable production and satisfies the readers and enthusiasts. Kindly readers are requested to send their valuable critics and suggestions to improve the book to k.kheiralipour@ilam.ac.ir or kamrankheiralipour@gmail.com.

Kamran Kheiralipour

Chapter 1

Production

1. Introduction

To talk about sustainable production, firstly production and production activities must be described. Hence, the purpose of this chapter is to provide the definitions, explanations, and descriptions of production activities for readers.

Each human act to produce a commodity or present service is named production activity. Production has high importance from economic, social, environmental, social, and political points of view.

In this chapter, the production statement was defined and its goal and importance were presented. Different production phases and output products were explained. Production factors and main aspects of production were described.

2. Definition

Production (Olsen, 2015; Fast, 2015; Akuchekian, 2017; Afshar, 2018; Selikoff, 2020) can be a simple or complicated activity. Production is the activity to present a usable product that can be sold at a price (Kotler et al., 2006). The production activities are done to satisfy needs, demands, or problems and have different aspects such as source, labor, technology, supply chain, market, energy, transportation, sailing, pricing, decision-making, and so on.

The production process may be the changing, correcting, and or combining of the inputs to create and or present a commodity or service. Also evaluating, assessing, improving, and optimizing a commodity or service can be named a production process.

Both commodity and service products are presented in different production sectors such as industry (Jürgens, 2000; Anonymous, 2009), agriculture (Ashraf et al., 2012; Anonymous, 2016; Ghosh, 2019; Gupta,

2020), food (Hui et al., 2007a-b) and natural resources (Wang, 2020), and others. Each sector relates to one or more academic fields.

Agriculture includes activities to produce plants (agronomic and horticultural crops) and livestock (animals). The agricultural processing process is a sector that produces food from agricultural crops and livestock. Industry includes manufacturing, mining, and utility activities to produce intermediate and final commodities from viewpoint of mechanical and materials engineering (Kalpakjian and Schmid, 2013), electrical engineering (Bernie, 1997), mine engineering (Clifford et al., 2020; Hartman, 2021), chemical engineering (Hartman, 2021), computer engineering (Lance, 2013), and so on. Other sectors mostly include the production of services in different fields such as basic sciences (Geographic, 2011), social sciences (Kuper and Kuper, 2004), medical and health (Noble, 2000), veterinary (Joanna and Bassert, 2021), art (Davies, 1991), and so on.

Production activities are divided into three divisions as primary, secondary, and tertiary phases (Fisher, 1939). Primary production is referred to extraction activities that are done in agriculture, forestry, fishery, mining, and oil extraction sectors to provide the raw materials. The secondary production phase includes manufacturing and processing activities for providing semi-finished or finished commodities from raw materials for the third phase of production. In the tertiary phase, the finished commodities are put in the hands of consumers such as distributive traders, banking, insurance, transport, communications, law, administration, education, health, and defense.

3. Production System

Production of a product is done by a system. Generally, a system refers to a set of interrelated and interacting parts which act according to the predetermined rules to form a unified whole (Backlund, 2000) and follows a specific goal.

The production system (Monden, 2011) is complex to produce a product. A production system includes labor, machinery, devices and so on that work together based on a discipline to produce a product. Maybe only one labor works in the small production systems and conducts all production tasks.

A system, surrounded and influenced by its environment, is described by its boundaries, structure, and purpose and expressed in its functioning. In big production systems, there may be different subsystems and many laborers work in each one.

Each production system includes input, process, output, and emission parts. Some inputs are used or consumed in each production activity and one or more outputs are obtained and one or more emissions are emitted to the environment by conducting specific processes (Figure 1.1).

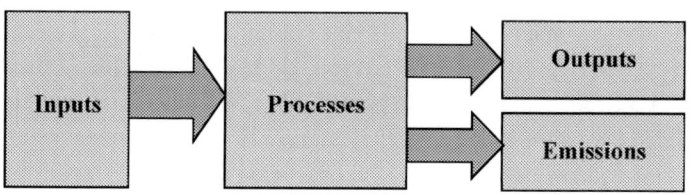

Figure 1.1. Inputs and outputs in production activities.

3.1. Input

Different factors in production activities are land, labor, capital, and entrepreneurship and which all play a distinctive role in production (Fisher, 2006). The inputs may be materials, energy, or immaterial (information, knowledge, and plan) that can be named production factors.

Labor refers to the human efforts for doing a job or presenting a service. This is the most important factor in production activities and needs special ability, skill, and experience.

Capital factor refers to production finance. It is required for providing facilitates to be used in production activities such as land or money to pay the labor wages, provide technology, and purchase the inputs to be consumed in the production process.

Entrepreneurship (Lumsdaine and Binks, 2006; Harrington, 2018) is the creation or extraction of value by using other production factors via initiatively and in some cases risky attempts. It includes designing, launching, and running a business to making a product.

3.2. Output

In some production processes, more than one output is obtained. In such cases, the outputs are classified into main products and byproducts (Muthu and Li, 2013) which are valuable and can be used.

A product can be an object or service, either physical, either virtual or cyber form. An object may be a raw material and may include a commodity, hardware, device, machine, and system or a part of them.

Service (Wirtz, 2017) is an activity to do something without producing a commodity so the services, unlike commodities, are not storable to be used in the future and are done simultaneously. Service includes conducting a process, transporting, distributing, and selling commodities, repairment and maintenance of a machine or system, and so on. Service needs the body ability, skill, experience, and ingenuity of persons or laborers to do that in the best manner. The person does a service manually or uses a machine, system, or software to do that.

Service refers to the production activity to do something without producing any physical commodities. In these activities, the customers use the body ability, experience, skill, and ingenuity of the sellers. Some service activities are done by laborers, only, whereas machines or systems are used in others. Washing and cleaning, selling, transportation and distribution of objects, repairment and maintenance of machines, automobiles, and systems, and so on are different kinds of services.

3.3. Emission

When running a production process, some pollutants may be emitted into the environment either air, soil, or water (Kheiralipour, 2020). The emissions are divided into two clusters, direct and indirect. Direct emissions happen when the inputs are used in the production process whereas indirect emissions are those emitted when producing the inputs in the corresponded production systems.

3.4. Process

Process (Weske, 2012) is a set of activities, tasks, and manners according to special methods and procedures to provide a product, either commodity or service.

The processes may be done by human labor directly or by devices and equipment (with a technology level) managed by professional humans. The production process may be a simple task done by a laborer such as transporting a commodity in business or a more complicated job done in an industry (company, factory, or organization) such as chemical, physical, electrical, mechanical, thermal, and so on.

A simple process is done manually by labor with or without applying any device to do service such as handling an object. In addition to humans, the complicated processes include devices, machines, systems, technologies, and or methods such as car production processes in a company.

The better production processes are those cause to obtain outputs with higher quality and quantity and lower inputs and emissions. With a constant amount of outputs, the lower the inputs' and emission values, the better the process. Also, can be told that the process is better which can produce higher outputs values with constant amounts of inputs and emissions.

4. Product Types

Products refer to both commodities and services (Garcia, 2014) because the production must have economic income.

4.1. Commodity

A commodity is a product that contains material or energy such as a car, machine, computer, building, bridge, road, fuel, animal, food, fruit, and so on.

Each product commodity has a certain life cycle. The product life cycle must be specified because it is necessary for analyzing production activities from the technical, environmental, and economic points of view.

Figure 1.2 shows the full life cycle for products. It includes different steps such as extracting, processing, using or consuming, recycling, reusing or reconsuming, and disposing of meterials. After disposal, the cycle is continued after a certain time, depending on the product kind.

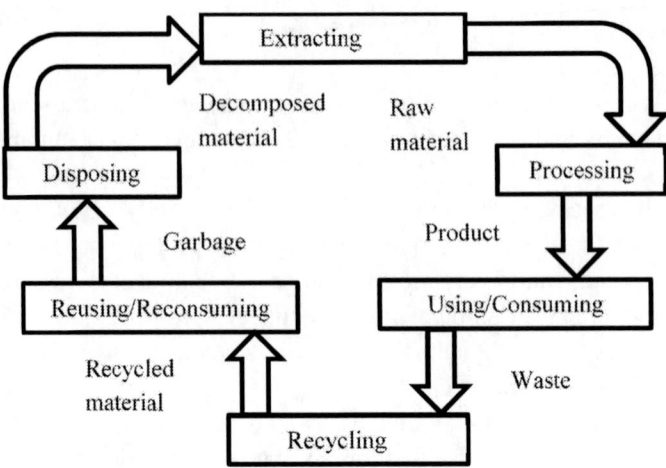

Figure 1.2. Product life cycle.

Production statement is used for each step in Figure 1.2 except for using and consumption, also using and consumption of inputs in a factory to produce a product are production activities.

Extracting process (Clifford et al., 2020) refers to taking out raw materials (Hulse, 2000) from an initial source, for example, bringing out iron ore from a mine. Also, all agriculture activities are a kind of extraction process because they obtain wheat grain from the land.

A process (Francis, 2015) is working on inputs to make a commodity or service product, for example, obtaining iron from iron ore, or tomato paste from tomato fruit.

Use and consumption are similar but the product is the same in its initial form after use (e.g., using water for cooling a hot system) whereas it differs from its initial form after consumption (e.g., consuming fuel in a car).

After using or consuming the products, some wastes may be observed. These waste materials can be used or consumed directly, without any change, by other consumers. This is called reuse or reconsumption (Nongpluh et al., 2013). If direct use of the wastes is not possible, they are recycled (Grohens et al., 2013) through a certain recycling process. Recycling refers to obtaining new materials by processing waste materials. The processes are done for materials after use or consumption by consumers in householders, companies, or factories. The obtained new materials are used or consumed by consumers, reusing/reconsuming.

Disposing of materials (Farraji et al., 2016) is putting unusable material in a landfill. These materials are decomposed after a certain time, depending on the material kind, and will become a source that can be taken out in extraction processes in the future.

4.2. Service

The service product is a kind of activity to do a task without producing a commodity such as repairing a car, machine, or computer, plowing a land farm, painting a wall, transporting a commodity, medical visiting a patient, teaching, testing, evaluation, selling, and so on.

All of the mentioned steps in the previous subsection are accompanied by some processes such as transporting, selling, purchasing, installing, repairing, maintaining, and so on, which are also taken into account as service products.

5. Production Types

A product can be produced in mass or batch production type, according to the product and organizational needs (Woollard and Emiliani, 2009). Mass production type refers to continuous manufacturing of the same commodity in large numbers. This type includes different workstations in which each one continually conducts a special task on the commodity one after another. Each station causes to more complete the commodity so that it is completed in the final station and is ready to be sold. In batch production same commodities are fully produced in a workstation only but producing the next product is started after the ending of the previous one.

6. Production Phases and Stages

Production has three phases including preproduction, during production, and postproduction phases (Oumano, 2011). The preproduction phase refers to operations that are done before starting production. During production means all operations to produce a product. Postproduction refers to operation after obtaining the outputs. These phases are different for different commodities and service products.

Production has two main stages, designing a design and implementation of the design (Kiran, 2018; Selikoff, 2020). These stages are involved in preproduction, during production, and postproduction phases. In any production activity or system, management is necessary to conduct the design and implementation stages. In this regard, production engineering (Radforld, 1980) refers to a field for proper planning and controlling the design, development, and operation of manufacturing systems via combining engineering, technology, and management. Designing and implementation, processes as well as management, are described in the next chapters.

7. Production Strategy

Strategy (Lawrence, 2013) is a long-term roadmap that includes a comprehensive plan to coordinate and organize actions to achieve overall goals.

Production strategy (Ivanov et al., 2017) refers to an overall plan which indicates the path to the goal of production activities. The first question that the production strategy must answer is which product must be produced? To answer this question, the strategy of production can be developed considering the people's needs versus imported products and also exported products versus the available resources in each city, region, or country. Another question that production strategy must answer is why a product must be produced, how a product can be produced, and so on.

8. Importance of Production

Production has high importance because it solves problems and satisfies the needs and demands of societies. It helps to conduct the tasks correctly, easily, and more reliable and safe. It decreases the time duration of the tasks because it helps to do them at a higher speed and consequently saves time.

Production activities create values and benefits from inputs such as raw materials. Also, it improves welfare because it presents more commodities which means more utility.

Production activities develop and improve the economy by generating income. This increases the economic power of the countries and consequently improves the political and defense power. Also, it helps to present humanitarian aid for poor peoples and countries and to provide more interference to improve the peace at the national and international levels.

Production develops and increases employment. Decreasing unemployment causes decrease social problems such as crime, disappointment, and lack of motivation in the societies.

Production increases the relations between peoples, societies, and countries. So it assists to enhance the fruition of relationship benefits such as familiarity with different cultures and so transmitting the positive beliefs and thoughts. So production improves and extends the culture and civilization.

9. Main Aspects of Production

The products are the designs or plans which have been implemented, manufactured, constructed, fabricated, built, or made. In addition to high productivity, the product must be practically usable, safe, reliable, competitive, and marketable. So the technical issue is one of the main important factors in production.

As production processes need inputs and each input need budget, any product has a production cost and consequently is sold at a price. The output has a higher value compared to the inputs, hence the product price must be higher than the production costs. So economy is one of the main aspects of production activities.

One of the main issues in production processes is emitting of pollution which is mentioned as emission. The emissions have environmental burdens and are dangerous to the environment. So one of the main aspects of production tasks is the environment.

The social aspects of production tasks are important because the customers have selecting ability. Besides quality and cost, interests, thinks, and beliefs (culture) of the consumers affect their choices. Hence, the social aspect must be taken into accounts one of the important aspects of production tasks.

Conclusion

Many of the activities in human life are production operations to produce a commodity or service product. Production activities empower the people, regions, and countries economically and decrease unemployment and other consequent social problems.

Production activities have different aspects like technical, economic, environmental, and social aspects. So these activities are important in each country. So, these aspects must be considered in each activity to produce the best products.

Chapter 2

Design

1. Introduction

Design is the first step in production activities and so it has high importance to have a better product. The design has a main role in sustainable production. This chapter describes the design and designing process. Definition of the design, design requirements, design importance, and design types are different contents of this chapter.

2. Definition

A design or plan (none) is a specification for producing a product, either commodities or services. A design may include drawings, calculations, and or programs to provide the general, special, technical, and engineering data, information, and knowledge to construct, manufacture, or fabricate a prototype object or implement an activity or process. Mechanical drawings, circuit diagrams, architectural blueprints, business processes, and clothing patterns are design examples. A design is a result of the design process.

Design, plan, or program (verb) (Gamma et al., 1995; Peters, 2005; Pahl et al., 2007; Tempelman et al., 2014; Ochsner and Altenbach, 2020) is an activity or process to formulate or develop a design, plan, or program to solve an existed or predicted problem or satisfy a specific need. Sometimes designing an operation is a decision-making process with or without any experiment (Budynas and Nisbett, 2014).

After designing a process, the design is considered to produce the expected product, commodity, or service, which must be functional, usable, manufacturable, remanufacturable, safe, reliable, competitive, and marketable. Sometimes there is no distinguishable boundary between designing and producing processes and a product is produced directly without considering any prior plan such as coding software or making a graphic design which is taken into accountas a design process. But, in some cases, the designing process includes research, negotiation, reflection, drawing, modeling, evaluation, interactive adjustment, and or redesign.

The goal of a designing process is creating an effect, satisfying a need, goal, and constraint, solving a problem through constructing an object, commodities, or system, or producing or implementing a servicing activity or process.

The designing activities have different aspects. It includes professional, social, environmental, and economic considerations.

Generally, the designer is applied to a professional person who produces a design. Specifically, different statements are used for those who work in a specific field such as engineers for electrical, mechanical, civil, specialty architects in the building design field, and others such as product designers, web designers, fashion designers, and interior designers.

Programing is a designing process because a program is a plan which is developed to execute a service. Program is a process for developing a schedule, network, flowchart, and instruction and also setting to be applied to do something. Sometimes, a person who develops a program is called a manager, whereas the programmer in computer sciences usually refers to a person who writes computer software.

3. Design Importance

In talking about the importance of the design, one can say all mentioned cases in the previous chapter about production importance. But, there is noted that the design is the first stage in product or service production. So, the designing process has higher importance compared to other production stages. Hence the design has a main role in sustainable production.

If the producers have a better design, they can produce a better product. It means that, without a good design, the producers maybe cannot produce a commodity product. This fact shows the importance of the designers' abilities. Besides the designer knowledge which is learned by studying the related books and other materials, the designing process needs mental, creative, and artistic abilities. Some levels of these abilities may be obtained by experiencing and thinking. Also, the designing tools have a main role in the designing process. Computer softwares have an important role in this regard because they have high importance as design tools.

4. Design Types

In one category, design types include innovative design, iterative or reverse engineering design, and the decision-making process. In another category, design can be divided into several items according to different fields. Here some design fields were presented:

- Design of mechanical parts (Shigley et al., 2004; Skakoon, J.G. 2008; Budynas and Nisbett, 2014; Childs, 2014; Ugural, 2020; Childs, 2021),
- Design of production line (Renna and Ambrico, 2021)
- Design of electrical and electronic devices (Sclater, 2003; Hauck, 2009; Thumann and Franz, 2009; Gift and Maundy, 2021),
- Design in civil engineering (Choi, 2004; Beeby and Narayanan, 2000; Addis, 2016)
- Architectural design (Ching and Eckler, 2012; Ingersoll and Kostof, 2012; Jefferis et al., 2016)
- Computer software or hardware design (Harris and Harris, 2012; Patterson and Hennessy, 2020),
- Agricultural machinery design (Bernacki et al., 1972; Varshney, 2004; Srivastava, 2005; Sharma and Mukesh, 2008)
- Product design (Baxter, 1995; Sule, 2007; Ulrich, 2011; Homburg et al., 2915; MacLean-Blevins, 2017)
- System design (Wasson, 2005)
- Clothing design (Baugh, 2011; Jamshidian et al., 2013; Lewis, 2015)

5. Designing Requirements

Designing is a complex operation and needs different aspects and tasks which must be properly managed. So, designing requires different knowledge, skills, and tools.

These attributes are necessary because they correctly affect the interest of the consumers to purchase the products and also the image of the producer brand.

5.1. Personal Abilities

The designing process requires some personal skills including, creativity, problem-solving, and communication in a corporation with the knowledge and information related to the general sciences including basic courses such as mathematics, statistics, computers, graphics, and national and international languages. Other main requirements of personal requirements are professional knowledge, specialized competence, and working with professional softwares in the specific field of design type.

The designing process includes some professional calculations. In universities, the designers learn the professional calculations of the designing process. Besides the above requirements, the designers must professionally work with high responsibility and cultivate a strong sense of professional ethics (Budynas and Nisbett, 2014).

5.2. Design Tools

To conduct professional calculations, drawings, programs, and statistical analysis, the designing process needs different computational methods and designing technologies as tools and software to do them correctly with low cost and time durations.

Recently, micro and macro computers assist designers to design a plan in all fields, called computer-aided engineering (CAE). For example, computer-aided engineering in mechanical engineering includes the required software in mechanical design as two-dimensional drawing and three-dimensional drawing or computer-aided design (CAD) software (SolidWorks, Unigraphics, AutoCAD, ProEngineer, Aries, CadKey, I-Deas, and so on), software for mechanical analysis based on finite element method and computational fluid dynamics analysis (ANSYS, ADAMS, Working Model MSC/NASTRAN, DADS, CATIA, Algor, CFD++, FLUENT, FIDAP, and so on), word processing (Quattro-Pro, Excel, Lotus, and so on), mathematical solvers (MATLAB, MathCad, Maple, 3 TKsolver, Mathematica, and so on), statistical analysis software (SAS, SPSS, and so on), and so on. So the designers must attempt to work with the tools correctly.

6. Design Stages

The design has main stages from start (need identifying) to end (need satisfying). It includes predesigning, design during production, post-production design, and redesign stages (Ullman, 2009).

6.1. Preproduction Design

Preproduction design is the first stage of design in various fields such as mechanics, electricity, civil, and so on. This main stage includes sub-stages as:

- Goal definition
- Analysis
- Research
- Target definition
- Problem-solving
- Evaluation
- Optimization
- Presentation

A goal (Locke and Latham, 1990; Rouillard, 2003) is a predetermined result to be achieved. It must be well defined by persons, companies, organizations, and countries due to its high importance and effects on the activities.

Design goal defining is the beginning step of the designing process. Usually, the first step or starting point of each work is hard, so defining the design goal is a very important step in the designing process. It means providing an early statement of design purpose based on the existing problem. In this step, the need, problem, or object must be correctly and exactly read, understood, refined, and identified, generally including the definition of the intended purpose in the design. The problem may include dissatisfaction, discomfort, or a malfunction or defective system. The purpose of the design may be to create a new device, machine, system, or a part of them.

The need may be identified and presented by a researcher, a product manager, a laborer, and so on, or recommended by a consumer. Some needs may be identified when analyzing a system by a systems analyzer person.

Analysis means assessing the design that if it can reach the goal. The process includes professional analysis, economic analysis, environmental analysis, and social analysis.

In the next step, research, search, or investigation is done to study similar designs and products. Research (Tan, 2017) is a creative and systematic scientific work that is done to find new knowledge that nobody know it, in other words, to extend the boundary of the existing knowledge in the world. Search is a studying, looking, and asking process to find the existing data, information, or knowledge that others know. The investigation is an attempt to detect a fact, information, or knowledge.

In the target definition step, all required attributes, features, and characteristics of the target to be designed must be defined and specified. This information includes inputs and outputs and their values, characteristics (dimensions, speed, accuracy, error, and so on.), and constraints (operating temperature, maximum range, expected changes in variables, weight, and so on).

In the problem-solving step (Eide et al., 2002), conceptualization and documentation of the design are provided. Concept (Keinonen and Takala, 2006; Perroux, 2010; Andreasen et al., 2015) is a mental thing to develop a design. A concept is a mental refined idea that is more exactly defined to receive a goal. Idea (Alexander and Wentz, 1996; Greiner, 1997; Paulin, 2014) is a seen rough image, notion, or impression in the mind, as a starting point of a concept.

After combining features and limitations, different mental designs should be proposed and quantified. Documentation means providing relevant technical drawings, sketches, and calculations of the design. Design documentation needs professional knowledge, skill, methods, and techniques of the designing process. These requirements are provided in the universities in different education sections such as machine component design, harvesting machine design, electrical circuit design, and so on.

In many cases, a model is provided. Modeling (Pidd, 1996; Haefner, 2005; Embley and Thalheim, 2011; Lohstroh et al., 2018; Prada et al., 2019) is providing a model to represent some or all characteristics, features, or behaviors of a product. This process may be data modeling (Reingruber and Gregory, 1994), mechanical modeling (Hurmuzlu and Nwokah, 2001; Pfeiffer and Bremer, 2017; Gouge and Michaleris, 2018), electrical modeling (Monthei, 1999; Li, 2012), and so on.

The final design or plan must be evaluated (Ross et al., 2004) to find if its performance may be satisfied before the presentation of that. If the design is unsatisfactory, it must be modified. If different designs are provided, it is needed to be compared for finding and choosing the better one. In this regard simulating is done. Simulating (Fanchi, 2005; Sokolowski and Banks, 2008; Popovici and Mosterman, 2017; Zeigler et al., 2018; Prada et al., 2019; Trine and Christini, 2020) is imitating the operation of a product to evaluate it by the examination of the real product or the representation of the working of its model.

For example, the finite element method (FEM) (Liu and Quek, 2003) as a numerical method and ANSYS Software can be applied for evaluating the designs in the field of mechanics, electricity, civil, biosystem, and so on. Using the software, besides improving the performance and efficiency, causes to decrease in time duration and cost of analysis (Jahanbakhshi et al., 2020).

After evaluation, if the design fails to meet the expectations, it needs to be optimized (Cavazzuti, 2013). Optimization means changing, correcting, or modifying the design for finding maximum output values when consuming constant values of the inputs or achieving minimum input values when obtaining constant output values. In the predesign process, optimization can be done by changing or correcting the concept, drawing, and calculations.

In the final step of predesign stage, the resulted designs or plans are presented to be implemented (made, fabricated, constructed, built, and executed). The presentation (Eissen and Steur, 2014) is an important step because it wants to indicate the effort of the previous steps. It needs communication skills for well presenting of the design so that can be understandable and acceptable.

6.2. Design During Production

The stage of design during production includes:

- Testing
- Developing

Testing and developing processes are conducted when the product of intense is produced based on its design. Testing and evaluating processes are the same processes (Ross et al., 2004). When producing, the incomplete products are tested in each progressing step to know if they will be satisfactory or not. If there is any mistake or problems, they must be corrected.

After testing, the product must be developed. Development (Ulrich, 2011; Kahn, 2012) of the product is growing the design to be better and its domain expanded.

6.3. Postproduction Design

The stage of postproduction design includes two steps:

- Implementation
- Evaluation

The goal of this stage is to provide useful feedback for future improvement and optimization of the design after manufacturing the product of interest.

After development, the final design is implemented to obtain the product, either commodities or services. The produced product in this step is called a prototype. A prototype (Blackwell and Manar, 2015) is not a final product. It is a produced product that must be tested before selling it. If it passed the tests, it is taken into account as the final product.

After the implementation step, the evaluation process is done to test the prototype product and based on the obtained results, some conclusions including constructive criticism, suggestions, and recommendations are provided for future improvement of the product. If the prototype passes the tests and evaluation, it can be sold.

6.4. Redesign

As the designing is an iterative process (Budynas and Nisbett, 2014) some or all stages and sub-stages of the process may be repeated to well satisfy the goal, problem, or need.

The final stage of the designing process is redesign. The goal of this stage is continually pursuing design improvements and productivity gains. Based on the provided conclusions in the previous stage, the whole design or product or one or more parts of them must be changed to be improved or optimized. Hence, one or more main and sub-stages of the designing process may be repeated to be corrected in order to present a better product to customers.

Conclusion

Design is the most important role in production activities to produce a commodity of service product because it is the first steps in production. The best design is required to have best product. So, the design requirements including designer properties and design tools must be considered to design a product. The design has different steps that all or some of them are necessary for the design process based on the product of interest.

Chapter 3

Implementation

1. Introduction

As mentioned in chapter 1, production has two main stages, design and implementation. The result of the designing process is a plan and that of implementation is a product, either commodity or service. These two stages are also taken into account as production activities. After the designing process, the resulted design, plan, or program is considered to be implemented to make a product. In this chapter, the implementation activity was defined and its requirements were explained.

2. Definition

Implementation (Harwood, 2013; Sandfort and Moulton, 2015; Coronel and Morris, 2018) refers to applying, executing, or using a design, plan, program, algorithm, policy, concept, or idea to produce a product, either commodity or service. This stage is done after designing a design and essentially must be properly managed. Also, the designing process needs management.

Implementation includes creating a commodity and or doing a service based on the prepared design. Some implementing types for creating a commodity are:

Implementing type	**Object**
• Assembling	• System
• Building	• Mansion
• Casting	• Metal part
• Changing	• Material Phase
• Constructing	• Structure
• Combining	• Alloys
• Converting	• Material
• Cooking	• Food

- Cultivating
- Cutting
- Fabricating
- Forging
- Forming
- Growing
- Jointing
- Lathing
- Making
- Manufacturing
- Molding
- Montaging
- Sewing
- Welding

- Crop
- Metal sheet
- Machine part
- Machine part
- Machine part
- Animal
- Mechanical link
- Machine part
- Food
- Car
- Machine part
- Car
- Clothing
- Machine part

Manufacturing (Thompson, 2007; Youssef et al., 2012) includes all required processes to produce a commodity from the start point to the finish point. Two main steps in the manufacturing process are construction (Merritt and Ricketts, 2000) and fabricating (Timings, 2008; Jeffus, 2011). Construction refers to forming an object and fabricating refers to creating a part of an object.

Some examples of implementing types to do service are:

Implementing type
- Adjusting
- Carrying out
- Conducting
- Driving
- Drawing
- Establishing
- Evaluating
- Executing
- Handling
- Harvesting
- Installing
- Leading
- Maintaining

Object
- System
- Duty
- Work
- Car
- Plan
- Organization
- Machine
- Law
- Commodity
- Crop
- Computer software
- Team
- Car

- Monitoring
- Operating
- Optimizing
- Planting
- Plowing
- Protecting
- Repairing
- Gathering
- Selling
- Setting up
- Storing
- Teaching
- Testing
- Training
- Translating
- Transporting
- Writing

- Machine
- System
- Machine
- Seed
- Land farm
- Environment
- Car
- Information
- Product
- System
- Crop
- Student
- Machine
- Robot
- Documentation
- Commodity
- Prescription

These implementation types are done according to their corresponding design. In another word, each product required one or more implementation processes corresponding to its design, plan, program, and so on. For producing a product, not only a design is necessary and must be properly designed, but also the corresponding implementation processes must be correctly conducted.

3. Implementation Requirements

The implementation process to produce a product based on a design, plan, program, and so on, requires some factors mentioned in this chapter. The requirements are infrastructure, technology, human, energy, and money which were explained in the next sections.

4. Infrastructure

Infrastructure (McDonald, 2001; Hayes, 2005; Hayes, 2016; Coffelt and Hendrickson, 2017) includes different physical materials which are required to produce a product. It refers to physical structures, facilities, and systems that are required to enable, sustain, or enhance societal living and working conditions in a company, factory, or society. It refers to roads, railways, bridges, tunnels, water supply, cooling and heating installation, sewers, electrical grids, audio and video equipment, telecommunications, internet, furnisher, and so on.

5. Technology

Technology (Chryssolouris, 2006; Nee, 2015; Brecher, 2015; Rajput, 2017) refers to both devices (Tlusty, 1999) and techniques (Lefteri, 2012; Davim, 2012) that are applied to do something in addition to infrastructures. Devices include hardware (tools, equipment, machines, instruments, utensils, and so on), software, and or the techniques include methods, skills, and processes to produce a product.

One of the main factors in the implementation of high-quality and high-marketable products is the available technology level in the production systems. Technology is divided into several categories, including low, medium, high, new, and emerging technology. Low technology is the old technology that has spread to large parts of human society, such as multimedia technology. High technology is one of those technologies that are taken into account as modern, complex, and expensive technologies. Medium technology is between advanced and simple technology. New technology is that one has just been produced and commercialized whereas emerging technology has not yet been fully commercialized and the full version has not been released (Rouhani and Keshavarz, 2014; Kheiralipour, 2021b).

6. Human Resources

Humans (Brauner and Ziefle, 2015; Agolla, 2018; Campbell, 2021) as staff and laborers are inevitably needed for the implementation of a design to produce a product. When simple and technical staffs do their tasks correctly, the design is implemented properly. This causes to make the expected product. So, training is one of the most important requirements in the implementation of a design. Human capital must have required characteristics (Pavon, 2010) were explained in the two next subsections. The general abilities of humans refer to those attributes that everybody must have, cultivate, and improve in life. In addition to general abilities, the implementation process needs professional abilities.

- Power (Olsen, 2017)
- Aptitude (Snow and Farr, 2021; Snow et al., 2021a-b)
- Honesty (Ciulla et al., 2013)
- Intelligence (Hunt, 2010; Mackintosh, 2011)
- Capability (Jaques and Cason, 1994)
- Worthiness (Manetti and Copenhaver, 2019)
- Potential (Vernon, 2009)
- Accuracy (Berkel et al., 2020)
- Knowledge (Geographic, 2009; Kheiralipour, 2021b).
- Skill (Brodner P. 1986; Arai et al., 2005)
- Responsibility (Branden, N. 1997; Fenwick, 2016; Alqahtani et al., 2021)

Power refers to the physical, mental, and spiritual ability to do the duty in a time duration with patience and without tiredness, and disgust.

One of the abilities of staff and laborers is aptitude and competency to do special work. Intelligence is another required ability. This ability is necessary because the staff and laborers must be logical, abstract thought, understanding, self-awareness, communication, and learning abilities and have emotional knowledge, retaining, planning, and problem-solving abilities.

Honesty refers to human truthfulness, righteousness, straightforwardness, loyalty, fealty, and avoiding bad traits such as lying, cheating, theft, cowardice, and oppression.

Capability means the inclusion and suitability of a person to be considered for a certain job.

Human worthiness is the merit and glory of staff and laborers. Each labor must know his/her right and respect the rights of his/her colleagues.

Human accuracy is an ability to do the duty correctly similar to the expected product and precision is doing the duty at different times similar to each other. Accuracy has high importance in working safely to avoid injury. The safety of the laborers and staff is one of the main aspects of the implementation step, because it allows for continuous production. In other words, the injured laborers and staff have lower power, accuracy, and patience to accomplish their duty. This causes frailer in producing the product. Also, substituting the injured person caused to lose the time, decrease the income, and increase the costs.

Appropriate implementation of a design needs well-trained staff and laborers, either simple or technical. All staff and laborers must do their duties safely, correctly, and on time. For that, it is necessary to learn the especial knowledge and skills as the specific abilities of the laborers and staff.

The workers must learn the special knowledge to do their duties. Knowledge is the human awareness obtained from information processing and information results from the processing of the data.

Also, they have to get the required technical skills to do their special tasks. Skill is the learned ability to practically carry out a task to obtain predetermined results.

Social responsibility means that each laborer must accept and do his/her duty and also form well cooperation with other laborers in a working group. Also, the laborers must admit their mistakes and attempt to compensate them. In addition to social responsibility, professional responsibility is another ability. It says that the laborers and staff must work properly in their specialty and professional manner. Also, they must obey the law, avoid conflicts of interest, and put the interests of consumers ahead of their interests.

7. Materials and Energy

To implement a design for obtaining a product, some inputs are needed as material and energy. These inputs must be prepared and properly consumed or used.

Materials (Tanchoco, 1994; Harris et al., 2003; Ptak and Smith, 2011; Kheiralipour and Sheikhi, 2021) are physical inputs that are consumed or used in implementing and producing a product. It includes metals, plastics, water, chemicals, soil, and so on. The materials contain hidden energy. In other words, the energy was consumed to produce them in the production factories.

Another input form in implementing a design is energy (Myers, 1975; Thiede, 2012) which refers to direct energy (Payandeh et al., 2016) such as fossil fuels (diesel, petrol, petroleum, and natural gas), renewable fuels (biodiesel, bio alcohol, and biogas), renewable energy (solar energy, wind energy, human labor energy, and so on). Energy is the power of doing a change or work. In other words, it is an essential input that must be transferred to a system to be changed or do a job. Different energy types are mechanical, electrical, chemical, and so on, which are divided into potential and kinetic.

In a production system, energy is divided into two forms, direct and indirect energy. Direct energy includes electricity, fuel, labor energy, and indirect energy are other inputs such as commodities that are not taken into account as energy but the energy was used for producing them in the producer system.

8. Money

Money (Graham and Zweig, 2006; Mishkin, 2007; Wilkinghoff, 2009; Michalowicz, 2017) is a verified record that is generally accepted as payment for purchasing the products and repayment of debts in a country.

Money or cash is a vital factor in implementing a design to produce a product so cash flow (Robert and Kiyosaki, 2011) is the lifeblood to calculate money entered into the activity and exhausted from that.

Money is necessary to purchase or rent the infrastructures and technology (fixed costs) and materials and energy inputs (variables costs). The money is needed to buy the materials, initial or recycled, and energies as fuels and human energy such as staff, laborers, and operators (as salary). Also, it is needed to meet new investments for developing business and producing new products.

Conclusion

Implementation is another step of production after design. In this step, the designed design or plan is used to produce a product. So, this step is important because to produce the best products, the best-designed designs or plans must be well implemented. Hence the requirements of the implementation process must be well considered by producers.

Chapter 4

Management

1. Introduction

Management is very important in executing an activity and needs to be properly done to reach a goal. It is needed for designing a plan and implementing the plan to produce a product. In this chapter, the definition of management was provided. Management requirements were briefly explained. Management production was described and all kinds of management which must be considered to successfully and continually produce a product were introduced.

2. Definition

Management (Brown, 1999; Morden, T. 2004; DuBrin, 2011; Pettinger, R. 2006) is the administration of self-affairs or a system or organization, either business or non-profit such as a governmental organization, office, company, or factory through the decision making and executing processes. The management consists of four main tasks including programming, organizing, leading, controlling, and supervising, which must be performed simultaneously and interrelated. Forecasting, setting, commanding, and coordinating are other duties of management. These duties are done by providing the strategies for the practical use of facilities and possibilities to reach the purpose of the system. In a system, the facilities and possibilities are human, land, knowledge, materials and energy, cost, machinery, equipment, technology, and so on. So, management is divided into the management of human resources, cost management, knowledge management, technology management, and so on.

A manager manages an organization, company, factory, or office. Based on the size of the organization, different managing levels may be defined as top, middle, and low manager levels. The duty of the top manager is to set the strategic purpose of the organization and make decisions on how the overall organization will be. A top manager may be a director board member, a chief executive officer, or a president of an organization that provides direction to the middle management level. At the middle management level, the strategic purpose of top management is communicated to the low management level. A middle manager may be a branch manager, regional manager, department manager, and section manager. Low managers are who oversee the work of regular employees and laborers and provide direction on their work and jobs. Team leaders, front-line managers, and supervisors are examples of low managers. In a small organization may be a person is the middle and low manager or do all management levels individually. In individual tasks, a person must be a manager and laborer to do the management and laboring duties, simultaneously. At all levels of management, and also for anyone with any job, self-management is an important issue.

3. Management Requirements

Besides general knowledge, management requires the professional sciences, and technical expertise such as management calculations and tools and software that are learned in the universities, courses, and workshops, and self-management skills. Self-management is setting, regulating, and controlling yourself in all aspects of your life such as feelings, thoughts, stresses, motivations, actions, works, and so on to do personal and job duties. Self-management skills focus on personal responsibility such as goal setting, time management, self-motivation, stress management, accountability, and organizational skills. Also, perception towards others, conflict resolution, resilience, patience, clear communication, and self-improvement are other self-management skills. These skills allow for improving human independence, ability to create opportunities, employability, productivity, functionality, and efficiency. Besides these skills, the management needs political, conceptual, interpersonal, leadership, behavioral, and diagnostic skills. Political skill is required to build a power base and to establish connections and communications with others. In management, conceptual skill is necessary to analyze complex, sudden, and or unexpected situations. Interpersonal skill is important to communicate, motivate, mentor, and delegate. The ability to visualize appropriate responses to a situation is called a diagnostic skill. Leadership ability is one of the main required skills to communicate a vision and inspire people to embrace that vision. Persuasive ability is one of the main abilities of successful managers. Also, a successful manager must be an honest, trustworthy, courageous, risk-taking, and eloquent person.

4. Production Management

Production management (Rasmussen, 2013; Crawford and Di Benedetto, 2014; Gupta and Starr, 2014; Khojasteh, 2018) includes planning, programming, overseeing, controlling, and optimizing the processes of producing an efficient product or service.

Besides accomplishing the product, the goal of the production management is to optimize resource use, reduce production costs, and improve the product quality, competitiveness, and brand. The duties of production management are using and controlling inventory such as designing, physical materials, machinery, equipment, methods, and money, supervising and training the laborers, and so on.

Production management means managing a production activity by applying techniques, methods, and manners to maximize net profit. The maximum profit is obtained through well introducing the products with the highest quality, the lowest cost, and so the highest marketability to the consumers. These ends are possible via proper management. Proper production management can be pursued by applying all or many management types (sometimes called business management types), depending on the size of the production system. Different management types are design, manufacturing, service, operations, engineering, human resource, program and project, knowledge, technology, public relations, supply chain, procurement, research and development, quality, risk, changes, innovation, facility, financial, marketing, sales, and strategic management. The management types have been introduced in the next subsections.

4.1. Design Management

Design management (Best, 2006; Emmitt, 2017) includes leading the design team in a production system. It focuses on converting an idea to a design to be used for producing a product. Design management has high importance because the designing process is the first step of production and the design quality greatly affects the net profit of the production system. Hence, the designing process must be properly managed to increase the design quality via continually improving the designs, plans, or programs.

4.2. Manufacturing Management

Manufacturing management (Gibson et al., 1995) means managing the activities to create goods and commodities. It includes construction management (Knutson et al., 2008) and fabricating management (Timings, 2008; Jeffus, 2011).

4.3. Service Management

The duty of service management (Normann, 2001; Haksever and Render, 2013; Bryson et al., 2020) is streamlining and managing the system's workflow to automate or support human decision-making. Also, service management attempts to understand the system services from the perspective of both the system and consumers to know the satisfaction level of the consumers. Knowing the satisfaction level of the consumers is necessary for the systems to manage the quality, costs, risks, and so on.

Service management is similar to information technology management; but the service management incorporates both automated systems and laborers.

4.4. Operations Management

Operation is a kind of activity that is applied to a commodity or service by humans, devices, or both. Operations management (Yoo and Glardon, 2018; Seubert and Vokey, 2020) is the responsibility for ensuring that all departments of business operations are efficient. Managing the operations of a business means dealing with a plethora of departments, strategies, and processes. Operations teams need to consider the acquisition, development, and utilization of resources their business needs to deliver the commodities and services clients want. Examples of operations management are maintenance management (Levitt, 2009), waste management (Sharma and Reddy, 2004), and so on.

4.5. Research and Development Management

Research and development (R&D) (Akhilesh, 2014; Daim et al., 2017) have an important role and are taken into account as key factors of progress because it is the first step of developing a new product, constructing or improving, by starting innovative activities.

Research and development management focuses on managing the research (basic or fundamental), developments (technologies, advances, concepts, new products, and processes), prototyping, R&D portfolio management, technology transfer, and so on.

4.6. Engineering Management

The duties of engineering management (Gupta, 2014) are similar to those of R&D management, but engineering management is more involved in manufacturing and constructing compared to R&D management. The engineering management more focuses on producing products based on the research results obtained by R&D management.

4.7. Innovation Management

Innovation refers to the implantation of a new idea to produce a new product or improve an existing product. Innovation management (Gilbert et al., 2018) is important to produce new products via analysis of the potential of innovation development. As innovation is related to different processes, innovation management usually cooperates with R&D, strategic, and change management.

4.8. Human Resource Management

Besides personnel administration, the duties of human resource management are applying employee programing and managing to make a positive impact on both the staff and the business. Human resource management (Bratton, J., Gold, 2007; Stredwick, 2014; Tyson, 2015) includes hiring, recruiting, compensation, safety and wellness, insurance and benefits, and encouragement and fines.

4.9. Strategic Management

Strategy is a plan to reach long-term purposes with high importance and includes setting the goals and priorities, determining the main activities to reach the goals, and mobilizing the limited resources to implement the activities (Freedman, 2013). Strategic management (Reed and Tech, 2020; Amason and Ward, 2020) focuses on the formulation and implementation of the main goals and initiatives in different technology, competition, marketing, financial, operational areas, and so on via managing the strategic thoughts.

4.10. Program and Project Management

Project management (Kogon et al., 2015; Layton et al., 2020) focuses on planning, executing, and supervising a project via prioritizing and obtaining the essential tools and knowledge to fulfill the needs of the project in both the short and long term. The duty of program management (Thiry, 2010; Brown, 2014) is similar to that of project management, but it involves more than one project.

4.11. Knowledge Management

Knowledge (Welbourne, 2001; Geographic, 2009) is theoretical or practical awareness, knowing, familiarity, or understanding something or someone. It includes descriptive (fact), procedural (skill), or acquaintance (objects).

Knowledge is divided into scientific and nonscientific knowledge (Pena, 1991; Yeatman, 1996; Falade and Coultas, 2017). Also, it can be divided into theoretical and partial knowledge. It is acquired by experience, education (perceiving, discovering, or learning), studying, searching, and researching. It is mentionable that knowledge differs from science. Scientific knowledge is obtained from information and the information is obtained from data by the scientific methods. Science provides the required explanations for obtaining knowledge. Science consists of the body of knowledge and the process of producing knowledge and includes systematic investigation (research, analysis, discussion, and interpretations) via scientific methods to obtain the knowledge.

Knowledge management (Dalkir, 2005; Becerra-Fernandez and Sabherwal, 2014; Massingham, 2019; Machado and Davim, 2021) refers to properly recording and using the available knowledge and obtaining new knowledge. The goal of knowledge management is to maximize the use and productivity of knowledge. It manages the knowledge of a person or knowledge in a system or organization via collecting or creating, classifying, storing, and distributing the data, information, and knowledge (Kheiralipour, 2021b).

4.12. Technology Management

Technology (Bolton W. 1988; Sharma, P. C. 2006; Jain, E.R.K. 2010; Dodds et al., 2021; Bukhman, 2021) has an important role in the advance and progress of the systems not only in point of technical, but also in economic, social, and environmental aspects. Techniques, skills, methods, and processes are called technology which are used (not consumed) to produce commodities or services, accomplish goals, or conduct scientific researchs. The duty of technology management (Gaynor, 1996; Kiran, 2008; Rastogi, 2009) is to assess the available technology in point of function, configuration, service, financial, and efficiency to purchase and use of them depends on the system needs. Also, evaluation of the role of technology in system progress is another technology management task (Kheiralipour, 2021a).

Information technology management (Liebowitz, 1998; Efraim and Carol, 2011) is a kind of technology management. Information technology is a kind of technology to design, plan, select, implement, use, and administer emerging and converging information and communications technologies.

4.13. Supply Chain Management

Supply chain management (Harland, 1996; Burt, 2011; Cimorelli, 2016; Stanton, 2020) focuses on the commodities and services flow from raw materials to consumers. It includes finding, providing, and storing raw materials, work-in-process inventory, and finished commodities as well as end-to-end order fulfillment from point of origin to point of consumption.

4.14. Procurement Management

Procurement management (Lysons, 2016; Van Weele, 2018) focuses on purchasing the needed products from external sources like markets, companies, and factories. Also, arranging the services from third-party providers is a duty of the procurement management. It is similar to supply chain management, but more than supply chain management focuses on budgetary limits and deadlines.

4.15. Facility Management

Facility management (Levitt, 2013; Hodges and Sekula, 2013; Roper and Payant, 2014; Jensen and van der Voordt, 2016; Ee, 2018) focuses on allocation, programing, and timing of the facilities to be safely used with high productivity and efficiency. Facilities refer to available assets that are purchased, built, installed, or established such as buildings, infrastructures, equipment, devices, machinery, energy, transportation, and furniture which support the systems to reach the their goals.

4.16. Financial and Capital Management

Financial management (Wilson, 2015; Gupta and Gupta, 2019) is finding an acceptable balance between profit and risk in a system to reach a profitable activity in the long term. Capital management (Tennent, 2014; Wilson, 2015; Brigham and Ehrhardt, 2016) is similar to financial management but for short time. Financial management relates to financial planning, financial control, and financial decision-making in accounting, investing, banking, insurance, securities, and other financial activities. Capital management relates to cash management, inventory management, and debtor management.

4.17. Marketing Management

The goal of marketing management (Kotler, 2002; Mullins and Walker, 2012; Hutt and Speh, 2012) is to increase the customer base, improve customer outlook and feedback, and increase the perceived value. Marketing management is a set of activities to develop branding opportunities and to practical execute marketing techniques, tactics, strategy, and pricing based on analysis of the system, collaborator, competitor, and customer analysis.

4.18. Sales Management

The activities in sales management (Noonan, 1998; Bellah, 2015; Weinberg et al., 2015; Brock, 2016; Ingram et al., 2019) relate to leading and overseeing sales teams. These activities are important to increase product sales by increasing the number of customers and their purchases. It wants to attract the interest of the possible customers and convert them to permanent customers.

4.19. Quality Management

Product quality is the first factor to establish a strong public image for a production system. Quality assurance is one of the most important insights in production systems because it has a critical role in maintaining the customers and their sales and satisfaction. The quality assurance tasks are monitored, controlled, and overseen by the quality management section.

Quality management (Gryna et al., 2005; Pyzdek and Keller, 2013; Rumane, 2019; Luthra et al., 2020) wants to maintain or increase product quality. The quality management section looks for the interests, demands, and critics of the customers and consumers and assesses them. Then, it submits some conclusions, recommendations, and suggestions to the design management section according to its assessing results.

4.20. Crisis and Risk Management

A crisis is an unexpected event, problem, or flaw which occurs at a specific point in time that damages the system. Crisis management includes crisis prevention, criss assessment, crisis handling, and crisis termination to save the system. In crises situations and possible risks, managing the changes in the systems can be helpful.

A risk is a potential event that may be happened and harm the system. The duty of risk management (Mollah et al., 2013) is the assessment of the system to identify problem areas or predict possible problems, defects, or flaws and attempt to remove or minimize them.

4.21. Change Management

Sometimes the system must be changed due to and depends on internal or external problems, conditions, and transitions. So, the systems must have high flexibility to be lasting via changing different system parts. Change management (Kotter, 2012; Dietrich et al., 2019; Dawson and Andriopoulos, 2021) includes leading changes in settings, programs, the team works, and so on.

4.22. Public Relations Management

Public relations (Lordan, 2003; Singh, 2011; Moss and DeSanto, 2012; Austin, 2015) has a finishing role in constructing a strong public image of the system. Public relations management must inform the public, primarily journalists, via establishing the enouncements and communications about with latest news, achievements, products, and so on. Public relations strategies may vary by industry, but they have a consistent end goal: a strong public image.

4.23. Energy Management

Energy management (Smith and Parmenter, 2015; Weber, 2018; Kumar et al., 2020; Kaya et al., 2021) focuses on providing energy balance via executing different programs such as energy audits to decrease energy use and increase energy productivity. Decreasing the use of fossil energy and increasing the utilization of renewable energy are other goals of energy management (Kheiralipour, 2021b).

4.24. Environmental Management

The environment is one of the main aspects of production and the production systems tend to produce and consumers tend to consume environmentally friendly products. The duty of environmental management (Theodore and Theodore, 2009; Ajith Sankar, 2015; Murali Krishna and Manickam, 2017; Oliveira et al., 2021) is to reduce environmental effects through designing and programing cleaner products, minimizing the use of natural resources, and minimizing emissions to air, soil, and water (Kheiralipour, 2020).

4.25. Recycling Management

Recycling is one of the main issues in production management. It focuses on the recovery of the used or waste materials to obtain new products. Recycling management (Tchobanoglous and Kreith 2002; Goodship, 2010; Rogoff, M.J. 2013; Subramanian, 2019) is programing recycling plans to reuse the materials in company and factory producers and householders for minimizing the emitting of pollutants, use of natural resources, and consequently production costs.

5. Importance of Production Management

Production management has not only of high importance, but also it is necessary because it causes to reach the goals with high productivity and safety, low use of natural resources, and low costs. It applies knowledge, technology, land, raw and recycled materials, and human resources properly by programming, planning, and designing. It increases employment and productivity and decreases unemployment and natural resource loss. Optimum utilization of scarce resources causes minimalized use of natural resources. Besides this, it attempts to decrease the environmental pollutants. Hence the production management establishes equilibrium in the environment. By producing high-quality products at low cost, efficient production management causes to decrease in product prices and so increases the welfare of people and improves living standards.

Conclusion

Each activity required the best management to be well done. In the same way, the management is very important in production activities to well execute the design and implementation process in producing a product. So the management requirements must be considered in production activities. To manage a production activity, all or a few different kinds of management must be considered to successfully and continually produce a product, based on the activity nature.

Chapter 5

Sustainability

1. Introduction

Sustainability is a hyper-disciplinary issue and becomes one of the main concerns in the world which is recently considered by researchers in various fields, producers, and consumers. It is more important due to high population growth.

Sustainability has no end point, like human prosperity. The more the prosperity, the better the human life. In fact, sustainability is a path that human beings must move on its road in order to improve the state of sustainability. In this chapter, sustainability and its aspects were defined and finally sustainable production was explained.

2. History of Sustainability

In ancient times, the sustainable term has been applied to people in Ilam, Iran. It was a kind of wishing for people to have continuous success. This term shows the importance of the appearance and reality of human activities and human success that must be continued in life.

Besides human, natural resources was are of the concerns of the people living in Palaeolithic time and some attempts had been done to preserve the soil fertility to continue farm activities, produce food for them, and maintaining resources for the future (Ike, 1984; Fukuyama, 2008).

In 1713, *Nachhaltigkeit* (sustainability in German) term was used in forestry to prevent wood harvesting more than new yields each year (Wiersum, 1995; Wilderer, 2007). Then, the economic aspects of sustainability were considered since 1798 due to resource limitations (Hotelling, 1931).

Finally, the sustainability concept with considering the environmental aspects was originated in 1987 by Brundtland Report (WCED, 1987). More information about the history of sustainability can be obtained by reading the books by Grober and Cunningham (2012), Caradonna (2014), and Caradonna (2018).

3. Definition of Sustainability

Sustainability (Kassel, 1750; Jacques, 2014; Caradonna and Ballerini, 2016; Clayton and Radcliffe, 2016; Walker, 2018; Brinkmann and Garren, 2018; Byrd and DeMates, 2019) is the capacity of enduring in a continuing way across different aspects of human life in recent and future (Kuhlman and Farrington, 2010). It is reached via proper use and/or consumption of human, material and energy resources, money, knowledge, and technology in an acceptable equilibrium to satisfy human needs in the short and long term.

Sustainability is reached by reorganizing living conditions, adjusting individual lifestyles, developing cleaner technologies, designing flexible and reversible systems, and developing green production. Eco-villages, eco-municipalities, and eco-cities are sustainable living conditions that can decrease the environmental burdens. Cleaner technology (Johansson, A. 1992; Kirkwood and Longley, 1995; Pernick and Wilder, 2007; Azad, 2020) is a kind of technology that can reduce the environmental burdens by improving the efficiency in energy consumption or production of processes, sustainable use of resources, and environmental protection activities. Cleaner technology includes green technologies, renewable energy producers, green transportation, electric motors, green chemistry, green lighting, green water, and recycling. Green production includes cleaner industry, green building, sustainable agriculture, sustainable architecture, and permaculture. The term "permaculture" (Holmgren, 2002) is an abbreviation of "permanent agriculture", that firstly used for land management to develop natural ecosystems based on comprehensive design principles, then for regenerative agriculture, and then for regenerative rewilding and community resilience inspired by Masanobu Fukuoka's natural agriculture.

The two main matters in sustainability are proper use of natural resources by the current generation and maintaining some of those for future generations. Sustainability wants to deal with the two matters as complement ones because the current generation uses natural resources and provides knowledge, technology, and welfare for both themselves and future generations. The sustainability concept originated from the interaction between the human seeking to have a better life and the constraint of resources (Kuhlman and Farrington, 2010). So, this end is accessible via considering different aspects of sustainability as economic, environmental, technical, and social (James, 2014; Magee et al., 2012) which were described in the next sections.

4. Sustainability Views

There are two views for sustainability as weak or economic and strong or ecological view (Stiglitz, J.E. 1997). Weak sustainability emphasizes compensating for the loss of natural resources by increased capital because the loss of natural resources is inevitable whereas strong sustainability emphasizes preserving natural resources because it is essential for human survival. The first idea points out that the next generation should inherit a stock of assets including human-made such as science, technology, infrastructure, landscapes, cultural heritage, and so on, no less than the stock inherited by the previous generation and the second tells that the next generation should inherit a stock of environmental assets no less than the stock inherited by the previous generation (Ayres et al., 1998).

The economic view of human life tends to cost-benefit analysis and wants to substitute natural resources with a human-made environment whereas the ecological view of human life tends to preserve the natural environment. Sustainability wants to solve the conflict between the two views of human life.

The two views can be applied depending on the resources kind. So, some resources are involved in weak variety and the remains in the strong sustainability. For example, fossil fuels are falling into weak sustainability because when they are finished, people can consume alternative renewable fuels instead of fossil fuels. An ongoing extinct species is fallen in strong sustainability because it cannot be recovered by the current state of knowledge and technology and so it must be preserved. So, the favor of the natural scientists is strong sustainability but economists tend to weak sustainability (Kuhlman and Farrington, 2010).

Two principal arguments for strong sustainability are ecosystem services and ethical stewardship (Kuhlman and Farrington, 2010). The argument for ecosystem services emphasizes those services in human life which supported by the natural ecosystem without any human-forced changes in the ecosystem (Potschin and Haines-Young, 2008). This is not possible in modern civilization. The argument of ethical stewardship emphasizes preserving the beauty and diversity of our natural environment for the current and future generations because it points out that humans are not allowed to destroy anything life. Based on this argument, any human activity must be assessed in point of its impact on the environment and anything inside it to estimate the value of the impacts, identify endurable and non-endurable impacts, decrease the impacts, and compensate for the impacts by improving the level of well-being (Tonn, 2007). So, this sustainability argument says what has to be sacrificed to achieve a certain increase in well-being (Kuhlman and Farrington, 2010), is it rationally compensable or not? If yes it can be done and if not it must be forbidden.

5. Assessing the Sustainability

Sustainability can be assessed for each product to identify its distance from real sustainability. In assessing sustainability, the product impacts must be estimated and interpreted on environmental, economic, and social aspects.

Both strong and weak sustainability views must be respected in impacts assessment. In this assessment, the weak sustainability view is applied to the substitutable environment which should be compensated whereas strong sustainability defines the thresholds for the non-substitutable environment and should not be crossed (Kuhlman and Farrington, 2010).

Sustainability assessment must provide necessary instructions to keep a balance between both sustainability views on one hand and human welfare on other hand.

6. Importance of Sustainability

The importance of sustainability is due to its key role in both human life and the environment, welfare for the first one, and conservation for the second one. Sustainability wants to provide permanent welfare for both present and future originations by considering all economic, social, and environmental aspects. Although real and sustainable well-being and happiness as the main goal of humans, depends on many factors such as freedom, religion, believes, thoughts, acts, and so on, sustainability can assist people to reach it. So sustainability is important for humans and all living binges in the world to reach justice.

7. Sustainable Development

Development refers to a new stage of changing situations of different human and natural acts and manners, such as personal development (Emerson and

Stewart, 2011; de Jong, E. 2014), which resulted in expanding, growing, and completing them.

Sustainable development was defined as a development that satisfies the needs of the current without destroying the possibility of satisfying the needs of future generations (WCED. 1987). So sustainable development emphasizes the welfare of both present and future generations via improving the life quality. Moreover, sustainable development includes comprehensive policies, programs, and or projects which cause higher human well-being and the lowest environmental impacts for the current and future generations (Ayres et al., 1998). This is possible by the development of the economic and social aspects of the people with preserving the environment (Elkington, 1994).

Sustainable development needs to apply sustainability in both production and consumption. Also, the two production and consumption terms are in line with each other. Consumption of inputs is done for the production of a commodity and the obtained commodity is consumed after production.

8. Sustainability Aspects

A simple example of unsustainable consumption is diabetes. This disease is caused in a person by consuming fat and sugar materials more than the amount the person needs. In these cases, resources (environmental aspects) and money (economic aspects) are lost on one hand and human health (social aspects) is endangered in another hand. Hence, different aspects of sustainability are important and related to each other.

Different aspects of sustainability are economic, environmental, technical, and social (James, 2014; Magee et al., 2012). On the sustainability path, all aspects of human life and activities must be designed, implemented, and managed in a sustainable way such that it includes human diet and food (Andersen and Kuhn, 2015; Richards, 2020) and other social aspects (Hammes, 2015; Benson and Craig, 2017; Minkoff-Zern, 2019), environment (Ayers, 2017) and climate (Robinson, 2018; Gates, B. 2021), production (Dadd-Redalia, 1994; Frick, 2016) and business (Carbo et al., 2017), and so on.

9. Sustainable Production

In sustainable production, the first goal of each production activity must be making a positive effect on the world, and obtaining economical incomes must be considered with other aspects of sustainability as environmental and social aspects. Also, in sustainable production, different stages of production, i.e., design and implementation, technical aspects must be properly managed for all activities, in addition to considering all economic and environmental aspects (Figure 5.1).

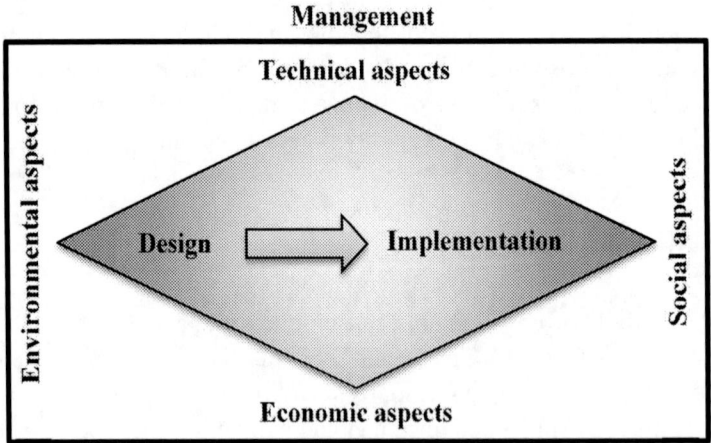

Figure 5.1. Sustainable production.

Sustainable production needs gathering and also invention required information to do and manage all steps of the activities, properly. In this regard, different books have been published on point of design (Ozolins, 2014; Frick, 2016; Walker et al., 2017; Ceschin and Gaziulusoy, 2019; Coady et al., 2020), seafood production (Bush and Oosterveer, 2020), industry (Reniers et al., 2013; Sahota, 2014), operations (Averill, 2011; Render et al., 2018), information and communication (Marolla, 2018), infrastructure (Gardoni, 2018), packaging (Jedlicka, 2008), civil (Braham, 2017), urban (Owen, D. 2009; Suzuki et al., 2013; van Maarseveen et al., 2019), business (Benna and Benna, 2018), finance (Silver, 2017), tourism (Fayos-Sola and Cooper, 2018), innovation (Fucks et al., 2015), entrepreneurship (Stenn, 2017; Perez-Uribe et al., 2018), policy (Shiva, 2005), strategy (Werbach, 2009; Rosenberg, M.

2015), energy (Van de Putte et al., 2017), management (Ulrich and Smallwood, 2013; Doppelt, 2009; Cohen, 2014; Friedman, 2020), and so on.

In each production activity to produce a commodity (produced in industry, agriculture, natural resources, and so on) or service (operation, management, policy, and so on), different elements must be respected in sustainable production. Different elements for sustainable production have been described in future chapters, Chapters 7-10. All or some of the elements must be considered in each activity.

Conclusion

Sustainability is a hyper-disciplinary issue that must be taken into account as a path in all scientific and research fields and production and consumption activities. It has high importance to reach an equilibrium in the world. Sustainability has different aspects that must be respected in all activities. Sustainable production is a new concept in the world to produce both commodity and service products in sustainability road.

Chapter 6

Economic Aspect

1. Introduction

The sustainable production activities must be robust in point of economical sustainability and those activities are continuable that their economic aspects are properly managed. So, one of the main aspects of sustainable production is economic. This chapter defines and explains the economic aspects of production activities which must be considered in both production steps as design and implementation of the designs. At first, the economy was defined and then different economical elements of production activities were explained. Economic elements are economic indexes such as price, cost, benefit, economic productivity, and economic efficiency.

2. Economy and Economics

An economy (Gregory and Stuart, 2013; CORE. 2019) refers to trade and exchange relations in different commodity and service production and consumption activities.

Economics (Dobson, 2004; Krugman and Wells, 2012; Sowell, 2014; Cairncross and Sinclair, 2014; Todaro, 2014) is an academic field that involves how people interact with wealth in the production, distribution, trade, and consumption of products, either commodity or services. Economics deals with different sections including microeconomics (Mas-Colell et al., 1995; Taylor et al., 2014), macroeconomics (Veseth, 1984; Thomas, 2021), international economics (Kenen, 2000; Krugman Paul et al., 2014), business economics (Harris, N. 2001; Dransfield, 2013), political economy (Sackrey et al., 2013; Balaam and Dillman, 2018), environmental economics (Hanley t al., 2019; Lewis and Tietenberg, 2020), engineering economics (Panneerselvam, 2013; Blank and Tarquin, 2017), agricultural economics (Drummond and Goodwin, 2010; Cramer et al., 2019; Barkley and Barkley, 2020), industrial economics (Devine et al., 1985; Clarke, 1991; Barthwal, 2010), social economics (Becker and Murphy, 2003; Murtagh, 2020), and so on.

3. Economic Analysis

Economic analysis in production and consumption activities means assessing the economical profits via cost-benefit analysis to decrease costs and increase incomes (Samuelson, 1983; Schumpeter, 1955; McAfee and Lewis, 2009; Rima, 2009).

Each input has a price which a cost must be spent to buy that (Wouters et al., 2012) and must be considered in different production steps, i.e., design process, and implementation of the design. It should be noted that the economic aspects of production activities should be considered simultaneously with other aspects of production such as environmental, social, and technical aspects due to economic analysis alone is not enough for sustainable production. The disadvantage of cost-benefit analysis is neglecting the future of the natural environment and also disability in the estimation of many essential variables related to that (Kuhlman and Farrington, 2010).

4. Economic Elements

Economic elements are those factors relate to the costs and incomes in production. The economical elements are important in point of sustainable economy in production activities and without that, the activities cannot be continued. Moreover, sometimes the economical elements influence environmental and social aspects in production activities. The economic elements are price, cost, benefit, economic productivity, and economic efficiency.

4.1. Price

A price (Mohammed and Murova, 2019; Armstrong, 2017; Simon and Fassnacht, 2019) is an amount of money that the buyer or customers must pay to sellers in return for one unit of product, either commodity or services. It is one of the main factors for the attraction of the customers because most the people buy the products at lower prices. So, pricing is an important strategy to choose a proper price for the products (Smith, 2011; Ingenbleek et al., 2013; Armstrong, G. 2017; Nagle and Müller, 2017).

Different factors affect a product's price such as production costs, supply conditions, demands, and so on. So, the product price must be defined properly and considered in designing and implementing processes of production. Besides them, the social and environmental aspects of production influence the price of the products.

4.2. Cost

A cost (Martin et al., 2007) is the amount of money that must be spent to produce a product, either commodity or service. Production cost includes both variable and fixed costs. So, all inputs and infrastructures must be considered to calculate the final cost of the produced products.

The production cost should be decreased as it as possible (Wheeler, 2010; Berk, 2010; Brundage et al., 2015). If the total cost of a product is high, it causes to increase in the product price and maybe decreases the customer attraction and so income. So, the total cost of the products must be calculated and considered in designing and implementing processes by producers. For that, variable cost (input costs) and fixed cost (infrastructure costs) must be calculated and summed, correctly. The spent monies to purchase the required inputs for producing a product must be recorded by the producers. Also, the share of monies to prepare and implement the infrastructures (especially with limited lives) in producing each product must be calculated and then added to the cost of the inputs. The total cost is calculated by the following equation (Barut et al., 2011; Nasseri, 2019):

$$TC = \sum_{i=0}^{n} N_i \times C_i \qquad (6.1)$$

Where TC is the total cost for producing a product, N_i is the number of the purchased input i, C_i is the cost of the product i, and n is the number of input types.

If a product has been properly designed and manufactured, the consumed and or lost material and energy can be decreased and this directly causes to decrease in the product cost. Also, it is important from the environmental point of view.

4.3. Gross Return

Gross return or gross income (Zangeneh et al., 2010; Komleh et al., 2011) is the total obtained money from selling the products. For different products with non-similar prices, the gross return is calculated by the following equation:

$$GR = \sum_{i=0}^{n} N_i \times P_i \qquad (6.2)$$

Where GR is gross return, N_i is the number of the sold products i, P_i is the price of the product i, and n is the number of product types.

4.4. Benefit

Benefit or net return (Mohammadi et al., 2008; Chen et al., 2020) is the net income that is obtained from selling a product. The benefit for one product is calculated by the following equation (Kitani, 1999):

$$NR = GR - TC \qquad (6.3)$$

Where NR is the benefit in the product production, GR is the gross income, and TC is the total cost of the product.

The economic benefit is increased by increasing the gross income and decreasing production costs. These must be considered in both design and implementation steps in production activities.

4.5. Economic Productivity

Economic productivity (Zangeneh et al., 2010; Ghorbani et al., 2011) is obtained when dividing the produced yield by the cost of production. It indicates the amount of obtained product corresponding to a unit of money for producing the product. It is obtained by the following equation (Kitani, 1999):

$$EP = \frac{Y}{TC} \qquad (6.4)$$

Where EP is economic productivity in the product production, Y is the amount of obtained yield, and TC is the total cost of the product.

The product design must be properly designed and implemented to increase economical productivity. It can be increased by decreasing the production costs. Also, the production process must be well designed and implemented to increase the yield.

4.6. Economic Efficiency

Economic efficiency (Cherchye et al., 2010; Sickles and Zelenyuk, 2019) or benefit/cost ratio is an indicator that shows the efficiency of investing to produce a product in production activities. Economic efficiency is calculated by dividing total benefits by the cost of product production. It is obtained by the following equation:

$$EE = \frac{NR}{TC} \tag{6.5}$$

Where *EE* is economic efficiency in the product production, *NR* is the benefit, and *TC* is the total cost of the product.

Economic efficiency indicates the benefit of investing a unit of money in the production activity. So, it has high importance in production activities and must be increased by decreasing production costs and increasing the net benefits. Although increasing the price cause to increase in the benefits, cost reduction is a better strategy because it causes to increase in customer interest.

Conclusion

The first aspect of sustainable production activities is the economic aspect. Because in sustainability path, those production activities can continually produce commodity or service products that have correct and robust economic aspects. The economic aspect of sustainable production activities has different elements. So all economic elements must be considered and managed in both steps in sustainable production activities i.e., designing of designs and plans and implementing them.

Chapter 7

Environmental Aspect

1. Introduction

One of the sustainability aspects is the environment which must be considered in sustainable production. Each sustainable production activity can be started after respecting environmental aspects. This condition must be applied for continuing the activities. In sustainable production, the activities must act in such a way that have the lowest impacts on the environment.

The environment was defined and its problems were explained in this chapter. Finally, the environmental elements including material usage, energy usage, environmental impacts, reuse, recycling, and disposal were discussed.

2. Environment

Environment (Bharuchs, 2004; Christensen, 2012; Nadakavukaren and Caravanos, 2020; Karr et al., 2021) is divided into the human, natural environment, and human-made environment. Human is the main part of the environment. So human safety (Anonymous. 1999) must be considered in production activities. The natural environment includes all living and non-living things occurring naturally whereas the human-made environment includes all things that originated from civilized human actions. Ecological units (Arihant, 2016; Anand, 2020) and naturally occurred phenomena (Wadhawan, 2017; Heidorn and Whitelaw, 2020) in their boundaries (soil, rocks, microorganisms, vegetation, atmosphere, rain, dust, flood, and so on) and universal natural resources (Holechek et al., 2002; Schellens and Gisladottir, 2018) and non-boundary physical phenomena (Sayigh, 1990; Sioshansi, 2011; Leal Filho and Surroop, 2018; Scorer, 2021) (air, water, climate, energy, radiation, electric charge, magnetism, and so on) are accounted as the natural environment.

Environmental sustainability wants to solve different environmental problems (Myers and Spoolman, 2014) such as environmental pollution (Markham, 2021; McKnight et al., 2021), environmental degradation (Shitole and Sable, 2012; Howard, 2020), climate change (Parry, 2021), and overconsumption (Reisch and Thogersen, 2015).

Different aspects of human activities as economic (Siebert, 2008; Cato, 2011; Hussen, 2019; Goodstein and Polasky, 2020), social (Pezzullo and Cox, 2021; Manfredo et al., 2014), and technical influences on the environment and vice versa.

3. Environmental Elements

Environmental sustainability (Michaelides, 2018) is one of the main aspects of sustainable production. The environmental aspects of sustainability include resource usage (material and energy), environmental impacts, reuse, recycling, and disposal elements.

The environmental burdens and impacts of the production systems must be assessed and decreased. Recently, assessing the environmental impacts of production systems is done by the life cycle assessment method (Kheiralipour, 2020; Kheiralipour, 2021a) and then the environmental impacts can be decreased by applying the assessing results in practice.

3.1. Environment Impacts

The equilibrium between the two views of sustainability, weak and strong sustainability, emphasizes impact assessment (Kuhlman and Farrington, 2010). Each activity has different effects on the environment (Gurjar et al., 2010; Harrison and Hester, 2012; Spellman, 2014; Payandeh et al. 2017; Roser and Ritchie, 2020; Kheiralipour, 2021a) which must be reduced. The first step in reducing the environmental impact of each activity is determining them (Kheiralipour, 2021a). In this regard, the life cycle assessment (LCA) methodology (Guinee, 2002; Lawrence, 2003; ISO 14044, 2006; Eccleston, 2011) is a standard, robust, and reliable tool that is applied to assess the environmental effects of different production systems to provide a comprehensive investigation in improving the environmental aspects. The method includes four phases including goal and scope defining, inventories analysis, environmental impacts calculation, and interpretation.

The first phase includes introducing the purpose of the LCA study, depicting and specifying the working plan and its subsystems, scope, functional unit, and reference flows. In the first phase, the boundary of the understudied production system is specified to specify the domain of the study, data, and results. The LCA goal involves different sectors as shown in Figure 7.1.

In the inventory analysis phase, the values of input and output materials and environmental emissions in the production systems are gathered and calculated. The emissions in the farms may be determined by standard methods and devices and using standard coefficients (Eggleston, 2006).

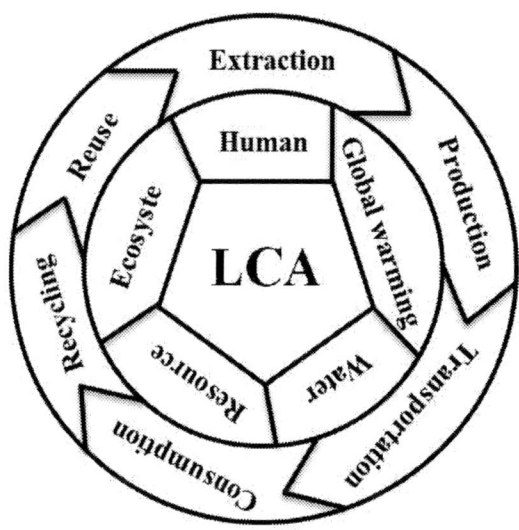

Figure 7.1. Life cycle assessment, goals, and subject activities (Kheiralipour, 202a).

In the impact assessment phase, the environmental indicators are calculated. In most cases, SimaPro Software from PRe Consultants (Consultants Pre, 2006) and CML-IA baseline V3.02/EU25/Characterization model are used to calculate the indicators. In this model, the value of eleven environmental impact groups including abiotic depletion, abiotic depletion (fossil fuels), global warming (GWP 100a), ozone layer depletion (ODP), human toxicity, freshwater aquatic ecotoxicity, marine aquatic ecotoxicity, terrestrial ecotoxicity, photochemical oxidation, acidification potential, and eutrophication are obtained.

The effects of inputs and emissions factors on the calculated impacts are calculated by the software in the fourth LCA phase. Also, some recommendations were presented in the interpretation phase (Kheiralipour et al. 2017a-b; Ramedani et al., 2019).

3.2. Pollution

One of the elements of sustainable production and environmental sustainability is environmental pollution (Markham, 2021; McKnight et al., 2021). Production activities may emit some environmental pollution to air, soil, and water (Kheiralipour, 2020). So the number of pollution must be reduced via decreasing the material and energy consumption, optimizing the processes, and decreasing the pollutants as well as recycling or reusing the emission.

3.3. Material Usage

Besides calculating the environmental impacts of production activities, another important element in sustainable production is material use (Kheiralipour and Sheikhi, 2020). The used and consumed materials are important in the case of sources used and economic costs and energy contents of the materials. Therefore, it is necessary to study the material flow in sustainable production activities.

Due to the high importance of material inputs in production systems, the type and amount of all inputs must be recorded. Also, the amounts of the output(s) must be determined. These data are useable to study the material flow and also calculate the input productivity indicator.

In the material flow study, the type and value of all inputs with their units are provided in a table. The productivity indicator shows the amount of the produced product, output, or yield for use or consumption of a unit of input material in the production systems. The following equation is used to calculate the productivity in the production systems:

$$P = Y/I \tag{7.1}$$

Where P is the productivity of a used or consumed input (unit of output/unit of input), Y is the produced yield or output (kg), and I is the considered input (m^3), either consumed or used. For example, this equation can be applied to calculate the water productivity in a production process. In this example, the equation determined how much yield is produced when consuming or using 1 m^3 water.

3.4. Energy Usage

The importance of energy in production is high because it is the main driver of work and change and has a vital role in sustainability. Energy is the ability to make a change in the form of work, heat, vibration, waves, and so on. Different kinds of energies from different sources (Skipka and Theodore, 2014; Nelson and Starcher, 2015) are used in production activities.

In addition to high costs and depletion of natural resources, energy consumption leads to emitting emissions (Kheiralipour and Sheikhi, 2020). Hence, energy is one of the main inputs in production processes. Energy analysis is done for decreasing energy consumption as one of the most important priorities in production to reduce consumption of sources, emission of pollutants, and economic costs.

The main stimulus for making work and any change in the world is energy. The energy issue is divided into two sectors as energy production and energy consumption. Energy production is the process of converting to energy by special systems like solar panels. Energy consumption is done by those systems that consume the energy to make a change or work. In both issues, efficiency is an important feature for both energy production and energy consumption. The increase in the energy efficiency of energy production systems increases the amount of the obtained energy. The enhancement of the energy efficiency cause to decrease in the energy loss in the energy consumption systems. Proper and optimal designs and management programs can assist to increase in energy production and decrease in energy consumption. Increasing energy production and reduction in energy consumption are the most important priorities in all activities to reduce the costs, depletion the natural resources (Kheiralipour and Sheikhi, 2020), and environmental impacts (Kheiralipour, 2020).

Energy analysis includes calculating the energy contents of the inputs and outputs, energy indicators, energy forms, and energy sources. The energy content of each input/output was obtained by multiplying the amount of consumption and production of each of them in the corresponding energy equivalent based on the following equation (Payandeh et al., 2017):

$$EC = I/O \times E \qquad (7.2)$$

Where E is the energy content of each input/output (MJ/Unit), I/O is the amount of input I or output O (Unit), and EC is the energy equivalent of each input/output (MJ).

Efficiency is one of the key indexes for the evaluation of energy production and consumption systems. The increase in the energy efficiency of energy production and consumption systems caused to increase in the amount of the obtained energy and a decrease the energy loss. Energy indicators are energy efficiency, energy productivity, energy intensity, and net energy gain (Jekayinfa and Bamgboye, 2008; Payandeh et al., 2016):

$$ER = \frac{OE}{IE} \tag{7.3}$$

$$EP = \frac{OY}{IE} \tag{7.4}$$

$$EI = \frac{IE}{OY} \tag{7.5}$$

$$NEG = (OE - IE)/OY) \tag{7.6}$$

Where ER is the energy efficiency, OE is the output energy (MJ), IE is the input energy (MJ), EP is the energy productivity (Unit/MJ), OY is the output yield (Unit), EI is the energy intensity (MJ/Unit), and NEG is the net energy gain (MJ/Unit). The designing and management processes to increase energy efficiency are well encouraged because they increase obtained energy in energy production systems and decrease energy losses in energy consumption systems.

Energy forms are calculated for inputs to determine the direct and indirect energy (Heidari et al., 2011; Zangeneh et al., 2010). Direct energies are those of the energy sources that are consumed (such as electricity and fuel) or used (labor) in the production systems; whereas the indirect ones are those that energy was used in the corresponding system, factory, or company to produce them.

Energy sources are divided into renewable and nonrenewable energy (Kheiralipour and Sheikhi, 2020). Renewable energies are come from renewable resources and naturally replenished. Renewable energy sources are sunlight, rivers, wind, rain, tides, waves, biomass, and geothermal heat. For example, sunlight is used to obtain solar energy and transform it into heat and electricity. Water in rivers, wind, and tides are used to obtain electricity or mechanical power. Biomasses are used to obtain biodiesel, bioethanol, and biogas for producing electricity or mechanical power. Nonrenewable energy has finite natural sources that cannot be replaced by nature in a short time.

Fossil fuels are nonrenewable sources including coal, petroleum, natural gas, and their derivatives such as diesel, petrol, grease, and so on. Other nonrenewable sources are earth minerals, metal ores, and groundwater in certain aquifers.

In point of energy sources, the goal is firstly to decrease the use of all energy forms, and then use renewable energies instead of nonrenewable ones as it is possible. So, the share of renewable energies is calculated for input used or consumed energy in the production systems. The following equation is used to calculate the share of renewable energy:

$$RES = RE/TIE \qquad (7.7)$$

Where *RES* is the share of a renewable energy source, *RE* is the value of the used or consumed renewable energy, and *TIE* is the total input energy.

3.5. Reuse and Recycling

After consuming a product, some substances may have remained which are called wastes or scraps. Waste or scrap materials may be usable or consumable. One of the main environmental elements is reconsumption or reuse (Grohens et al., 2021) because it causes to decrease in material usage from the initial resources. For reuse or reconsumption, sometimes inspection, cleaning, repair, or/and recuperate are needed. For example, the packages of the incoming parts for a product are reused for packaging the outgoing product or a part of the product itself.

Recycling (Waite, 2013; Worrell and Reuter, 2014) is conducting a reprocessing step of materials to produce new commodities. It may be applied for materials after use by consumers or may be done on waste materials in companies and factories.

Recycling is an environmentally friendly process that causes to decrease in the usage of raw materials. Through this, the resources are more conserved and protected. As producing raw materials has high costs due to extracting and production, using recyclable material causes to decrease in costs of the final commodities and consequently increases the income and economic benefits.

The commodities must be designed and constructed so that one can easily separate their components to be better recycled. The materials can be recycled that do not pollute the final commodities. Steel, aluminum, plastics, papers, woods, and agricultural materials are examples of recyclable materials. One kind of recycling is energy recovery which converts material waste to an energy source such as heat or biomass.

3.6. Garbage Disposal

Garbage is a type of scrap or waste material that cannot be reused, reconsumed, or recycled. Garbage disposal (Daniel, 1993; Vitorio Andreoli et al., 2007; Epstein, 2015) is the incineration process of garbage or placing them in a sanitary landfill. These processes are unavoidable for some materials to reduce their negative effects on the environment.

In sustainable production activities, disposal must be done in such a way that it has minimum negative effects on the environment. So the design and implementation steps in the activities must be properly done to reduce the disposal effects on the environment.

Conclusion

The environment is one of the main sustainable production aspects because we have borrowed it from new generations. Each sustainable production activity can be started and continued after respecting the environmental aspect. So optimal use of all natural resources and protection of them are important in both steps of production activities as design and implementation. In this regard, the environmental elements must be considered in each activity to produce a commodity or service product with the lowest environmental impacts.

Chapter 8

Social Aspect

1. Introduction

Social aspects in production activities have important roles in sustainable production because the social aspects of customers affect their interests and so purchasing the products. So social aspects must be considered in the design and implementation steps of the production activities and also the products must be well designed and implemented to have positive effects on the social aspects of human beings. In this chapter social aspects of sustainable production have been discussed. Different elements of sustainable production from the point of view of social aspects have been explained.

2. Social Domain

Social (Kuper and Kuper, 2009) refers to the collectively live of human populations that relate and communicate interactively. Social sciences focus on human beings and their respective relationships through the mentioned fields.

Social is related to the public and is in contrast with the private phase of human life. The customer element is described as technical in Chapter 10. It refers to the private phase of human life, but social contrasts with that because it is related to the public phase of human life.

The social domain has different fields. Besides economic, environ-mental, technical, and political fields, the geography, history, civics, and sociology fields affect the social aspects of human life. Although each of them is a separate field, they were mentioned as the social domain in this chapter. Culture, psychology, philosophy, and anthropology are incorporated into the mentioned fields. These fields have been explained in the next section as social elements.

3. Social Elements

To increase the marketability of a product, to be accepted and purchased by the customers, the social elements must be considered when the product is designed and implemented. Also, the products must be well designed and implemented to have positive effects on the social elements of human life (Kimball, 1933; Cocklin, 1995; Sasvari, 2012; Gurr, 2018). So the social elements are important in sustainable production. Sociology, geography, history, psychology, philosophy, anthropology, civics, and culture have been introduced in this section as social elements.

3.1. Culture

Culture (Tylor, 1871; Alic et al., 1989; Inglis, 2005; Kuper and Kuper, 2009; Pahlavan, 2009; Bartold and Suhrawardy, 2010; Hellemans, 2017; Zouelm, 2018; Mandour, 2019; Storey, 2021) is a broad and complex concept that affects all components of human life that its root is human thought. In fact, culture is human thought and material and nonmaterial based manners of humans which they formed according to that, consciously or unconsciously.

Culture is which involves different aspects of human societies as thoughts, attitudes, desires, interests, behavior, religion, beliefs, ritual, ethics, knowledge, art and music, architecture and shelter, capabilities, habits, customs, literature, clothing, cooking, technique, and technology of people.

Culture is passed down from the current generation to future generations and also it is changed and improved and new cultures are formed through individual and public learning.

Culture strongly affects sales and marketing because it forms customer interests and desires. So, knowing the culture of the goal customers and consumers is important for producers to design, manufacture, implement, and provide the products that meet their culture.

3.2. Sociology

Sociology (Giddens et al. 2007; Zandvakili, 2010; Little, 2016; Salvatore, 2016; Griffiths et al., 2019) refers to different aspects of human society including social behavior, relationships, and interactions. Also, different aspects of human culture associated with everyday life are considered in sociology. It includes social stratification, social mobility, religion, law, health, welfare, economy, military, punishment and control, education, and social capital. Sociology science focuses on social order and changes studies.

Different aspects of sociology mainly affect production. The needs of the societies with higher welfare and economic power are different from those of the poor people and those with low welfare. So studying the social aspects of the goal customers by the producers is vital to be familiar with consumer problems and needs. This has an important role in the economic aspects of sustainability.

3.3. Geography

Geography (Whitbeck, 1934; Jovanovic, M. 2001; Jovanovic, M. 2012; Bjelland et al., 2017) refers to the surface of earth and lands and their physical features, inhabitants, phenomena, and the relationships of humans with them.

Geography is such important that also the name of the places may affect the interests of the customers. Geography is important in point of selling products because the attributes of spaces and places affect health, climate, plants, animals, and economics. For example, some features of a special product which has been designed for a cold area may differ from that designed for a warm region. The requirements of the product for consumption in regions with higher humidity are different from those produced in dry areas. Hence the producer must study the geography of the goal customers and design and implement the products according to that.

3.4. History

History (Arnold, 2000; Bourdaghs, 2014; Claus and Marriott, 2017) is the events that formed in the past. The events may be cultural, natural, social, economic, and political. History science refers to the discovery, record, collection, organization, presentation, analysis, and interpretation of events.

The history of a city, region, and country influences people's national and international relationships, policies, interests, desires, and so on. Hence gathering information about the history of the society of the goal customers is important to produce the products to satisfy their needs, problems, and interests.

3.5. Civics

Civics (Bimber, 2000; Desjardins and Schuller, 2006; Brunold, A. 2015; Moradian, 2018; May and Ross, 2018) is the rights, obligations, and behavior of humans that affect other citizens in society. The civic study focuses on the theoretical, political, and practical aspects of relationships, rights, and duties of citizenship.

If a product is in contrast to the civics of a country, it cannot be sold easily in the goal region. Hence, civic must be considered in designing, manufacturing, and distributing the products.

3.6. Psychology

Psychology (Koffka, 1962; Fernald, 2008; Vahab, 2010; Coovert and Thompson, 2014; Rosen et al., 2015) focuses on the mind and thought and conscious and unconscious behavior and feelings of individuals and human society.

Behavioral and mental processes of humans include personality, perception, emotion, attention, motivation, cognition, intelligence, subjective experiences, brain functioning, interpersonal relationships, resilience, human development, and so on. These factors influence human needs, problems, and interests. Producers can consider these elements and their factors to produce the required products for each society.

3.7. Philosophy

Philosophy (Sellars, 1963; Nuttall, 2002; van de Poel, 2020) focuses on the general and fundamental questions, critical discussion, and rational argument related to existence and reality, reason, knowledge, belief, ethics, inference, values, mind, and language. The areas of philosophy are all fields of study but it more relates to religion, mathematics, natural science, education, and political fields.

Philosophy is associated with human wisdom and intellect and also culture. Consequently, it can affect society's needs and problems that can be solved by producers. So it has an important role in sustainable production.

3.8. Anthropology

Anthropology (Pfaffenberger, 1992; Hylland Eriksen, 2004; Suchman, 2011; Bryant and Knight, 2019; Squires, 2021) focuses on human biology, behavior, culture, language, society, and art in the present and past originations. Sometimes archaeology is included in anthropology studies. Absolutely anthropology affects human requirements. For example, human biology influences the needs of the people in each society. So, the producers must be respected by anthropology in designing and manufacturing the products to meet the needs of the goal society.

3.9. Policy and Politics

Policy (Dye, 1976; Fischer et al., 2007; Blakemore, 2013; Hill and Varone, 2021) includes a set of intentional guidelines which is implemented in form of a procedure or protocol in political and nonpolitical (managerial, financial, and administrative) mechanisms to guide decisions and reach the logical outcomes. Policy leads the organizational programs, decisions, and spending via identification and choosing significances, priorities, and alternatives.

Policy differs from strategy (Chapter 2). The policy is planned for a certain time and focuses on optimal use of the existing conditions and getting the best feedback.

The presidential executive and parliamentary orders are governmental policies. The policies are governed by the public sector including companies, factories, corporate, groups, and individuals.

Policies can be adopted by individuals, companies, organizations, societies, and governments. The governments can make subjective or objective decisions to reach a balance between energy use and economic incomes and between environmental preservation and incomes economic.

Political science is a branch of social science that studies politics and government. Politics (Connolly, 1981; Baak, 2000; Jevons, 2008; Colomer, 2010; Glass and Rose-Redwood, 2015) is a set of decisions of individuals or groups to achieve any form of power such as relations, resources, status, and so on. It determines the level of peace, negotiations, conflicts, and cooperation between the political parties. Also, the laws are made by politics and it determines the level of laws' enforcement.

Politics includes internal and international policies. Internal politics is exercised on a wide range of national levels including tribes and clans, parties, markets, companies, factories, institutions, and government. International politics is the same as the internal one but at the international level.

3.10. Law

Law (Willis, 1926; Burton, S.J. 1995; Kelsen, 2005; Harris, 2016; Hage and Akkermans, 2014) includes a set of governmental and nongovernmental legislations which created and enforced to regulate behavior and implement justice. Governmental laws are created and implemented by governmental institutions at national and international levels. The nongovernmental laws are those deified by companies, factories, and corporates to be implemented inside them to satisfy the justice between their staff and laborers and between the two of them at national and international levels. The laws can be created by a single or group of legislators including a judge and or executive.

Law differs from the policy. Policy guides the actions to achieve the desired outcome without compelling, forcing, prohibiting, or penalty whereas law is a set of constitutions, decrees, regulations, rules, statutes, or legally binding to compel, force, prohibit, or penalty humans in return to their behaviors. For example, taxes law is defined to get taxes on income from producers.

An activity or product which is in agreement with the law is legal and consequently illegal refers to a thing that is against the law or breaks the rules. Producers must be attended to law to produce the legal products. If it is not respected, the consumed energy and materials in producing a product are lost because it cannot be sold, and also they have to endure punishment and so economic losses.

Conclusion

The designs and plans must be well designed and implemented so that the resulted commodity and service products have positive effects on the social aspect of human beings. Also, the social elements are important because the products must be produced based on the people's interests and needs. So, the social elements must be considered to produce products in both design and implementation steps in sustainable production activities.

Chapter 9

Technical Aspect

1. Introduction

This chapter discusses the technical elements of sustainable production. To produce commodity and/or service products in sustainable production, technical elements must be well respected. The technical elements are necessary to produce products with high quality and marketability. Moreover, the elements must be in line with economic, social, and environmental elements in sustainable production.

2. Technique

The technique or technical aspects of production (National Research Council, 1990; Buchanan, 1998; Semeradova and Weinlich, 2020) refer to professional knowledge about the calculations, considerations, operations, processes, the technology of a specific field that is required to produce a product, either commodity or service.

The technical aspects of production are very important to produce products with high quality and marketability. Also, it is important in sustainability because it can optimize and decrease the material and energy use in production and so decrease the environmental impacts. Hence technical aspects of producing commodities and services must be respected in sustainable production.

3. Technical Elements

Technical elements (Childs, 2013) in design and production are process, production facility, production constraint, quantity, quality, timescale, standard specification, safety, liability, reliability, performance, aesthetics, mass, size, ergonomic, installation, maintenance, product lifespan, life in service, storage, shelf life, test, documentation, packing, packaging, liability, shipping, brand, competition, and market. The elements have been described in this section. In sustainable production, all or some of the required technical elements must be respected to produce the best environmental-friendly products with high incomes with positive effects on society.

3.1. Process

Production is done by conducting one or more processes on the inputs. The production process (Klocke, 2009; Groover, 2011; Sharma, 2016) is a set of actions such as changing, combining, and so on to produce a product. It includes different techniques, methods, and manners. There are two kinds of processes, natural and human-made. The natural processes are done naturally without human intervention whereas the human-made processes are conducted by humans, machines, and or systems.

The efficient production processes (Sujova et al., 2017; Cezarina and Frumusanu, 2017) are those which produce higher quality products with lower time duration, material loss, energy consumption, and cost. Skilled laborers with high work conscience are important for better conducting of the processes.

3.2. Production Facility

Production facilities (Sule, 2008; Kadane and Bhatwadekar, 2011; Greene, 2011; Perez-Gosende et al., 2021) include general and specific infrastructures and technologies to produce a product. Production facilities provide the possibility of doing a task and cause to do it easier and faster with higher accuracy. General infrastructures refer to roods, buildings, and all fixtures located therein such as installation, furniture, and so on, and specific infrastructures include machinery, technology, equipment, and so on.

Building and structures are important and must be well prepared based on the related standards because each production system needs specific structures in point of size, safety, and so on. More efficient technologies cause the production of better products with lower time duration, material loss, energy consumption, and cost.

3.3. Production Constraints

Production constraints (Hutchin, 2002; Nkomo et al., 2021) are particular limitations that must be considered or satisfied in any product production stage, from designing a design to consuming the produced products. The constraint is temporary or permanent. Constraints may be technical, financial, environmental, political, legal, and so on.

Sometimes the constraints forced the producers to change the quality, quantity, or price of the products. This role of constraints emphasizes on the high flexibility of production systems.

3.4. Quantity

Quantity (Knight and Kotschevar, 2000; Bali, 2011; Chase, 2020) is a magnitude property of the products. Sometimes increasing product quantity causes to increase in net income because in these cases the efficiency of consuming the inputs such as humans, materials, and energy is improved and so the net production income is increased.

3.5. Quality

Quality (Gitlow, 2000; Jameel et al., 2015; Nanda, 2016; Pristavka et al., 2016; Ling and Mansori, 2018) is an attribute to illustrate a product's superiority or state of being. Quality is a perceptual, subjective, and conditional property and sometimes it is perceived differently by different people.

Quality is the first factor for highlighting and improving a producer's brand image because it is the first consumers' reason for purchasing the product and accordingly compares the products with competitors in the marketplace. Product quality firstly refers to the strength of the product against frailer, wear, corrosion, and so on. Other factors which show the quality of the products are material, process, performance, accuracy, efficiency, reliability, safety, life, storage, shape, surface, and so on. So the producers must consider different elements to produce high-quality products. The above factors must be respected during the manufacturing process, but some factors must be considered after the manufacturing process.

In the case of agriculture production activities, some factors must be considered after harvesting the crops. Sometimes high-quality product refers to those agricultural crops without internal or external defects (Kheiralipour et al., 2015 & 2016; Farokhzad et al., 2020), and sometimes refers to crops with uniform ripeness and color (Mohammadi et al., 2015), shape (Kheiralipour and Pormah, 2016), size, volume, and mass (Mirzaee Ghaleh, 2008) and sometimes it refers to pure product without impurities (Khazaee et al., 2021). So grading, sorting, separating, and cleaning operations (Kheiralipour et al., 2018) are necessary for agriculture and similar production activities to increase product quality.

Production of the products with low quality causes to decrease in marketability, the number of customers, and income. Also, it causes to lose the materials because unpurchased commodities must be discarded or in the best condition recycled. In such cases, energy and costs of the first production are lost and so in point of economic and environmental aspects, it is the bad end.

3.6. Timescale

Timescale (Williams and Albertson, 2005; Kuehn, 2015; Spasojevic Brkic et al., 2016; Mahesh et al., 2018) is a scale applied to evaluate the time duration of a process. The time duration of a production process must be decreased because it causes reduction in energy consumption and also the product can be presented to markets faster. In some cases, reducing the time duration avoids product deterioration.

Timescale shows a process to be conducted in a specific time duration with a standard delay.

$$TS = AT + TD \tag{9.1}$$

Where TM is the timescale, AT is the allowable the time and TD is the time delay. The unit of timescale can be second, minute, hour, day, or year depending on the task.

3.7. Standard Specification

Standard specification (Khan and Abdul Raouf, 2005; ASTM International, 2012) is a set of documents including necessary physical and functional characteristics information about the product parts, size, tolerance, materials, processes, and so on which must be considered in designing, manufacturing, and implementing stages to produce a product with high quality, performance, efficiency, uniformity, and safety.

Depending on product types, the standards specifications are provided by government agencies, trade associations, corporations, and standards organizations such as SAE, AWS, NIST, ASTM, ISO, CEN, DoD, and so on). Some Standard specifications are national and or international ones that must be considered by all nations.

There is important to consumers that the purchased product has been produced based on the specified standards because they agree to buy the product upon all national or international requirements (Blake and Bly, 1993).

3.8. Safety

Safety is the state of being protected from danger, hazard, or risk. The dangers are not the same in different areas, sometimes the danger is harm exerted on humans, animals, fruits, and vegetables, sometimes it is a mechanical failure, and sometimes it is an economic loss. Here, safety (Budynas and Nisbett, 2014) refers to sureness against failure (crack, break, scab, wear, and so on) which is stated by the safety factor or design factor.

Safety is the main factor in product quality. The products must be well designed and implemented to reach a high safety factor to avoid early failure and malfunction.

3.9. Liability

Liability (Mildred, M. 2001; Zenios and Ziemba, 2007; Campbell, 2007; Daynard and Legresley, 2012; Koziol et al., 2018) refers to the responsibility of the producers for any damage and harm to the consumer caused by their products, either commodity or service, due to their defects.

In addition to satisfying the customer's needs, the manufacturers must have a high level of liability and take the responsibility for their work. Therefore, the designer must take responsibility for the final product. Professional ethics must be observed. To start on the path to success, the designer must see the necessary training and learn the necessary skills and professional ethics before starting work. This matter shows the importance of design and implementation stages in production more than other elements of sustainable production. In this regard, evaluating, testing, and quality control are necessary and important tasks in production.

Liability of the producers caused to present guarantee and warrantee (Murthy and Jack, 2003; Murthy and Blischke, 2006). The guarantee is a promise of producers to consumers that the product will be repaired, replaced or the money replayed if the product is below the specified quality and a warranty is a written assurance to show the truth in the commodities and it will be repaired or replaced if they are not.

3.10. Reliability

Generally, reliability (IEEE, 1990; Pham, 2003; Misra, 2008; Birolini, 2010; Kapur and Pecht, 2014) is an overall consistency of an attribute. In production activities, reliability or availability is used for commodities, especially, for devices and equipment, and refers to the ability of the commodity to work without failure in specified conditions for a time duration.

Reliability is stated as a digit between 0.00 (high error) and 1.00 (no error) or presented with a percent unit between 0.00 to 100.00%. It is obtained after conducting a set of tests on a group of the same systems and shows the value of the random errors. For example reliability of 0.90 means that there is a 90 percent chance that the product can safely work without failure. The producers must design and implement the design to achieve specific reliability. A system is highly reliable if the same results in testing or using are obtained, repeatedly. High-reliability score (close to 1.00) shows high precision, reproducibility, and consistency of the system in repeated tests.

In point of reliability, the product producers can go on the sustainability rood by improving the reliability of the products and presenting the correct reliability scores for them. The first strategy can be reached by conducting accurate tests and collecting information from the consumers and finding the conditions, causes, and reasons for failures. Presenting the correct level of reliability for produced products by producers has high importance for consumers to accurately manage the product usage with high functionality or prevent or predict the product failure.

3.11. Performance

Performance (Cooper, 1984; Cooper and Kleinschmidt,1995; Mcdavid et al., 2020) refers to carrying out or accomplishing an expected function. Efficiency is emerged to show the ability of a system for carrying out its performance. Efficiency is stated as a coefficient between 0.00 and 1.00 or with a percent unit between 0.00 and 100.00.

For example, a diesel engine produces mechanical power from diesel fuel (Khoobbakht et al., 2019), and the engine performance is delivering mechanical energy. In this case, how much diesel fuel energy is converted to mechanical energy by the engine? More efficient engines gave a higher amount of mechanical energy from the same amount of diesel fuel. So, proper design and manufacturing of an engine, cause to decrease in the usage of natural resources for producing the same power output and is better economically and environmentally.

3.12. Beauty

The appearance beauty of products (either goods or services) is in the realm of aesthetics (Taliaferro, 2011; Parker, 2012; Nanay, 2019). Various factors such as shape, surface, color, size, softness, and elegance show the beauty of the product.

Beauty is the first element that attracts consumers because it appears in the eyes of the customer. Therefore, it is very important in product marketability and causes to increase in income.

3.13. Material Type

Material type (Ashby, 2011; Mehmood et al., 2018. Benavides-Serrano et al., 2019) affects the quality, production cost, mass, and size of the produced products. So the material types are important in sustainable production and must be selected properly.

3.14. Mass

Mass is a property of a physical body with the unit of the kilogram (kg). Also, it refers to a measure of body resistance to acceleration when a force is exerted to that. Sometimes, weight is used instead of mass, it is incorrect because weight (with the unit of N) is the mass value multiplied by acceleration gravity (9.81 N/kg).

Mass and weight (Pine et al., 1999; Chaudhry et al., 2014; Jiang et al., 2015; Zhu et al., 2018) are related to the amount of material used in producing a product. The designing process must be done to avoid any extra mass, otherwise, more material and resources are consumed and cause to increase in the environmental burdens and also the production cost.

3.15. Size

Size (Cassandra and Josephson, 2009; Radojicic et al., 2013) refers to the length, width, and thickness dimensions of products. These dimensions determine their diameter, perimeter, area, and volume. It is divided into two kinds as single dimensions such as length, width, height, diameter, perimeter, and multiple dimensions (area and volume).

Same as mass and weight, size is related to the material used to produce a commodity product. So it has an important role in the economic and environmental aspects of production. Hence the size of the products must be correctly designed and considered in the manufacturing process.

3.16. Ergonomic

Ergonomics or human factors (Kroemer and Grandjean, 1997; Bridger, 2003; Wilson and Corlett, 2005; Dul and Weerdmeester, 2008; Stanton et al., 2014; Berlin and Adams, 2017; Kroemer and Kroemer, 2016; Ray and Maiti, 2018) refers to considering of psychological and physiological principles of human in production a product. When a product is designed and manufactured, the psychological and physiological aspects of its consumer or operator must be applied.

Ergonomics focus on the interaction between humans as the main member of the environment and products. It deals with decreasing the psychological and physiological comfortingly under the effects of various factors such as dimensional non-fittingly, harm, vibration, sound, noise, dust, and poison on a human to increase human comfort, safety, productivity and decrease or eliminate the human tire, error, illness, and injury which may be caused by them.

3.17. Installation

Some products have different parts or components. In these cases, installation (Ming, 2000; Gupta, 2019) means an assembly of the different parts to prepare the expected product. This production step must be correctly done to achieve the expected performance of the product.

3.18. Maintenance

Some products need maintenance and repairment (Mobley, 2002; Duncan and Richardson, 2014; Mobley, 2014; Ben-Daya et al., 2016). Maintenance of a product is a process started at the purchasing point of that and is continued to the discarding point of the product. The process is necessary to maintain the performance of the products such as equipment, machinery, buildings, and systems. It includes functional checking, servicing, and repairing of necessary parts of the product. Repairment means the recovery of a product such as machinery by correcting, changing, or replacing its malfunctioned parts. Servicing is applied for cars, automobiles, and so on and refers to providing engine oil, water, and other requirements of the engines.

3.19. Product Lifespan

Product lifespan or lifetime (Heiskanen, 1996; Shinsuke et al., 2010) is the total time of product life. It is the time from the point of purchasing and to the point of discarding of product. Product lifespan differs from product economic life because product economic life includes the time between purchasing point and the point that maintaining the product is more expensive than replacing it.

Also, product lifespan differs from product technical life and functional life. Product technical life refers to the maximum period of time (Cooper, 2010) that a product has the physical capacity to function. Functional life (Cox et al., 2013) refers to the time that a product should last without any external intervention to increase its lifespan.

3.20. Life in Service

Life in service (Cooper, 1994; Evans and Cooper, 2010) is the total lifetime of a product that is serviceable. This factor must be considered in the designing stage of a product production because it is an important role in the material used to produce it. Life in service differs from the product lifespan. Life in service includes the time duration that a product is effectively used whereas product lifespan refers to all times from purchasing to the discarding of the product. Life in service must be specified by the producers because it is important for consumers to manage the product service and replacement and predict the purchasing time of the product.

Life in service shows the service longevity of a product so consumers have to consider the performance, quality, and costs. If a product is used more than its life in service, its performance, efficiency, and productivity are reduced and the repairment and maintenance cost is increased.

3.21. Storage

Storage is used for a vast variety of things including commodities (Williams and Wright, 1991), foods (Shearlock, 2019), clouds (Yang and Xiaohua, 2014), energy (Marco et al., 2020), and so on. It means putting a commodity in the corresponded store such as market, silo, fridge, battery, and so on with a specified storing condition.

Each kind of commodity needs its corresponded store and storing condition. If a commodity is not stored in the required store or in unsatisfied condition, it is corrupted or malfunctioned faster which means a loss of resources and money. So the producers must correctly define the store and storing conditions for each commodity and consumers respected those correspondingly.

3.22. Shelf Life

Shelf life or storage time (Man and Jones, 1994; Corradini, 2018; Galanakis, 2019) refers to the maximum time duration which a commodity can be maintained in the corresponded storage without losing its function. It refers to the time duration that a commodity can be sold, consumed, or used after storage.

The shelf life is applied for commodity products, not services, such as medicines, medical devices, pharmaceutical drugs, foods and beverages, chemicals, tires, batteries, explosives, and other perishable items. It is informed to consumers by the expiration date. So the producers must emphasize the correct carving of the expiration date on the commodities because it has an important role in human health and avoids loss of resources and money.

3.23. Test

Test or evaluation (Raj et al., 2002; Mehta et al., 2005; Prasad and Nair, 2011; Hellier, 2012; Mehta et al., 2016) is a process to determine the performance, accuracy, quality, merit, worth, significance, or any other features of a product such as material, machines, devices, equipment, systems, and so on. The tests may be conducted according to a set of standards.

In evaluating a product, some attributes may be not expressed in quantitative terms in their nature, although they are important. These features may be left out only from the measurement, not from the assessment.

The tests are done by the producers before selling the product. The consumers must be informed about the results of the tests. Also, the test process is important to avoid selling the undesired products because selling the undesired products causes to decrease the marketability and income.

3.24. Documentation

Documentation (Schubert, 2000; Seufert, 2014) refers to required information and instructions about the produced product such as its parts, assembly, installation, use, and maintenance. This information is necessary to better usage and maintenance of the product. So the producers must present correct information in the product documentation.

Previously documentation has been provided as hard-copy papers called catalogs. Now, besides hard copy, the documentation can be found on online webpages and analog or digital media such as an audio tape or digital such as a CD as help, user guidelines, or user manual.

3.25. Packing

Packing (Gerald and Friedman, 2010) is the safe enclosing of the packaged or unpackaged commodities into bigger boxes, cartons, and so on to protect them against damage during the transportation process. Packing is all of the materials used to brace, contain, and protect a shipment.

A pack is a shipper box, carton, or container that contains packaged or unpackaged shipments. Packing takes place for shipping before packaging to a packaging unit or is done as a final process after packaging for preparing the commodities to transport them to the markets.

The role of packing in environmental sustainability is decreasing or avoiding material and energy losses. As protecting is the goal of packing, it prevents the damage of the packed commodities and consequently, decreases the commodity's wastes. As material and energy are used to produce the commodities, the packing prevents the material and energy losses. Decreasing material and energy losses cause to decrease in economic loss.

3.26. Packaging

Packaging and packing are not quite the same but sometimes they are called instead of each other as synonyms. Packaging (Blackwell, 2000; Emblem and Emblem, 2012; Kumar Agariya et al., 2012; Farmer, 2013; Ahmed et al., 2014; Mersid Poturak, 2014; Morris, 2016; Barros-Velazquez, 2016; Garcia-Arca et al., 2017) is the science and technology of putting the commodities in a box, container, or jar to be transported, distributed, warehoused, sold, and used. It includes designing, producing, using, evaluating, and optimizing processes of packaging containers. A packaging system refers to a system that puts or pours the commodities into a container is transported, stored, sold, and end-use. Also packing refers to all of the materials used to pack the commodities.

Packaging is used to protect the packaged commodities against damage in the markets or when transferring (after packing) such as a machine part or a set of them, a biscuit, a chocolate bar, and so on.

Packaging is used to package solid bulks (grains, legumes, beans, and so on), liquid (mineral water, milk, juices, and so on), and semisolid (jam, yogurt, and so on) materials for sale in the markets, and so on. In some cases, the packaged meat, fruit, and vegetables are sold in the markets. In these cases, a packaged commodity is the smallest sellable package.

The vital goal of packaging is preserving the perishable, spoilable, putrescible materials against corruption, spoilage, and putrefaction, such as jam, fruit juice, and so on. For fragile commodities, various products are used to surround the commodities inside the packs to protect them against moving around, mechanical damage, and breakage such as bubble wrap, corrugated cards, shredded paper, and foam cushioning. Sometimes packaging provides the required air or pressure around the materials inside the package. For many commodities, the packaging is used to seal them against water, dust, and pollutants. One of the strategic applications of packaging is sales increasing. Packaging is one of the main procedures to present commodities in an appealing way to attract customers and so increase the incomes. Because the commodities which are presented in beautiful packaging catch the consumer's eyes and sell better than others. Also, the packaging prevents pilferage via using the special packing material. Besides these, the packaging is used for advertising, marketing, and labeling. The commodities information is presented by packaging such as price, manufacturer, and representative images.

Packaging development strongly depends on the commodities. As packaging is applied for displaying the commodities to make them visually appealing for the customers, the appearance beauty of that is very important. Other aspects of packaging development are structural design, performance, cost and price, marketing, shelf life, quality assurance, logistics, legal, regulatory, graphic, resources and recycling, end-use, environment, etc.

Same as packing, packaging has an important role in environmental sustainability. Besides avoiding damage, packaging decreases the material corruption of the packaged commodities, it decreases the commodities wastes and so the material and energy losses are reduced.

Packaging is an augmented task that improves the value of packaged commodities. Also, by preventing material and energy losses, the packaging is useful in economical sustainability because if it is not used, the wasted commodities cause to decrease in income through increasing financial loss, the decreasing attraction of the consumers, and depreciation of the brand image.

3.27. Shipping

After producing and packing/packaging the products, they are transported to be received by distribution centers, retailer markets, or consumers, to be supplied as stocks for a future production process, stored in storage, and or moved to a building. These tasks are managed called the shipping process.

Shipping (Fricker and Whitford, 2004; Meyer, 2016; Colvin, 2017) refers to organizing, preparing, documenting, executing, and reporting the transportation process of commodities. Different ways are existence for the physical shipping of the products. Common ways to move the commodities are road, rail, sea, and air. Special ways are inland waterways, pipelines, and cable, especially for shipping the energy sources.

The shipping process must be properly managed from sending information to receiving the commodity. The product information must be correct and in time sent to the buyers. Selecting proper transporting way, safe and on-time loading of the commodities, and monitoring the transporting process and its security to avoid damages and corruption are important in the shipping of products.

Besides these important issues, the product and its pack and package must be efficiently designed according to the shipping distances, conditions, and so on to avoid failure or corruption. Also implementing the design for manufacturing the product must be properly done accordingly. These tasks have an important role in sustainability because they can avoid product failure or corruption and so avoid material and energy losses.

3.28. Brand

Brand (Ghodeswar, 2008; Sammut-Bonnici, 2015; Durmaz, Y., Yasar, 2016; Silva Chaves, 2017) is a name, term, design, symbol, or any other feature to distinguish one producer from others. It causes to create some values called brand equity. For the same products, the more famous brands, the higher the producer incomes. The brand attracts more purchasers and consumer and so increases the income. Hence brand is important in point of economy.

Various factors affect a producer brand. The first factors are related to product quality and beauty. So the design and manufacturing of the products are important in branding. Hence the brand is important in point of environment because proper manufacturing of the products avoids product malfunction and corruption and so material and energy losses. Other factors are related to the proper design of the brand.

3.29. Competition

Competition (Alchian and Allen, 1983; Tang, 2017; Hudson, 202) is an attempt conducted by a producer to produce better products for attracting more customers compared to other producers and consequently obtain more income.

There is mentionable that truth and ethics must be respected in competition. The competition is a desired attempt for producing higher quality products and if it is not possible, cheaper products with the same quality.

The existence or possible rivals of a produced product must be identified by the producer and it is considered in different production steps such as designing, manufacturing, selling, and so on to attract the customers. If the competition is not considered, the products may not meet the customers' interest for a long time and they are malfunctioned or corrupted. This means material and energy loss and economic and environmental drawbacks.

3.30. Market

A market (Furrer, 2006; Armstrong, 2017) is a place where sellers sell products, either commodities or services, to buyers. Sellers and buyers meet each other face-to-face in physical markets whereas the buyers purchase the products face-to-face communications in virtual markets. Marketability is the ability of products, either commodities or services, to attract customers and easily can be sold with acceptable price and profit. A product with higher marketability is more bought by consumers compared to a product with lower marketability. So this element has an efficient role in increasing the producer income.

Marketability (Murthy et al., 1994; Knotts t al., 2009) is a complicated element because different factors affect that such as price, newness, appearance, quality, brand, and so on. Hence it must be considered in designing and implementing the products by producers.

Another factor that affects marketability is customer interest and the products must meet that. A customer (Courage and Baxter, 2005; Thilmany et al., 2008; Forbes, 2016; Xiao, 2019) is a user or consumer that receives the products, either commodity or service, from a seller, either producer or supplier. Customer is important because there is no income without customers. So, the products must be designed and implemented based on the customers. Also some information about the product advantages must be provided by sellers to attract customers.

Conclusion

The technical aspect of sustainable production includes many elements. Almost all elements are necessary for the design and implementation steps to produce high quality commodity and service products with high marketability. For example, all of the elements must be considered for producing a mechanical commodity; but few of them are not required for service products such as storage, packaging, and son. Also, the technical elements must be in line with economic, social, and environmental elements in sustainable production.

References

Addis, W. 2016. *Structural and Civil Engineering Design.* Routledge. Oxfordshire, UK.
Afshar, A. 2018. *How to Properly Produce: Product Management in Starting and Developing a Start-up Business.* 1st Ed. Nasle Roshan Publication, Tehran, Iran.
Agolla, F.J. 2018. *Human Capital in the Smart Manufacturing and Industry. Digital Transformation in Smart Manufacturing.* 1st Ed. IntechOpen. London, UK.
Ahmed, R.R., Parmar, V., Amin, M.A. 2014. Impact of product packaging on consumer's buying behavior. *European Journal of Scientific Research* 120 (2): 145-157.
Ajith Sankar, R.N. 2015. *Environmental Management.* Oxford higher education. Oxford, UK.
Akhilesh, K.B. 2014. *R&D Management.* 1st Ed. Springer. Delhi, India.
Akuchekian, M. 2017. *Creating Lean Product-How to make a supreme world-class product?* 1st Ed. Dibagaran Publications, Tehran, Iran.
Alchian, A.A., Allen, W.R. 1983. Exchange and Production: Competition, Coordination, and Control. 1st Ed. Wadsworth Pub Co. Belmont, California, US.Armstrong, G. 2017. *Principles of Marketing.* 17th Ed. Pearson. New York City, New York, US.
Alexander, N.J., Wentz, A.C. 1996. *Idea to Product: The Process.* 1st Ed. Springer-Verlag New York, Us.
Alic, J., Kraft, M., Vig, N.J. 1989. *Technology and Culture. Technology and Politics.* 30(4): 1082.
Alqahtani, A.Y., Kongar, E., Pochampally, K.K., Gupta, S.M. 2021. *Responsible Manufacturing.* 1st Ed. CRC Press. New Jersey, US.
Amason, A.C., Ward, A. 2020 *Strategic Management: From Theory to Practice.* 2nd Ed. Routledge. Oxfordshire, UK.
Anand, V. 2020. *Environment and Ecology.* 1st Ed. McGraw Hill Education. New York City, New York, US.
Andersen, K., Kuhn, K. 2015. *The Sustainability Secret: Rethinking Our Diet to Transform the World.* 1st Ed. Earth Aware Editions. Totnes, UK.
Andreasen, M.M., Hansen, C.T., Cash, P. 2015. *Conceptual Design: Interpretations, Mindset and Models.* 1st Ed. Springer International Publishing. Aargau, Switzerland.
Anonymous. 2016. *Agricultural Crops Production: Beginner.* 1st Ed. 3G E-learning LLC. New York, US.
Anonymous. 2009. *Mass Production, Products from Phaidon Design Classics.* 1st Ed. Phaidon Press. London, United Kingdom.
Anonymous. 1999. *Safety and health in agriculture.* International Labour Organization. Geneva, Switzerland.

References

Arai, E., Kimura, F., Shirase, K. 2005. Knowledge and Skill Chains in Engineering and Manufacturing. In: *Information Infrastructure in the Era of Global Communications.* 1st Ed. Springer- Verlag. New York, US.

Arihant, E. 2016. *Environment and Ecology.* Arihant Publication India Limited. New Delhi, Delhi, India.

Armstrong, G. 2017. *Principles of Marketing.* 17th Ed. Pearson. New York, US.

Armstrong, T.J. 2008, Chapter 10: *Allowances, Localized Fatigue, Musculoskeletal Disorders, and Biomechanics.*

Arnold, J.H. 2000. *History: A Very Short Introduction.* New York: Oxford University Press. Oxford, UK.

Ashraf, M., Öztürk, M., Ahmad, M.S.A., Aksoy, A. 2012. *Crop Production for Agricultural Improvement.* 1st Ed. Springer Netherlands.

Asdrubali, F., Desideri, U. 2018. *Handbook of Energy Efficiency in Buildings: A Life Cycle Approach.* Butterworth-Heinemann. Massachusetts, US.

Ashby, M.F. 2011. *Materials Selection in Mechanical Design.* Butterworth-Heinemann. Oxford, UK.

ASTM International. 2012. *Form and Style of Standards, ASTM Blue Book.* The American Society for Testing and Materials (ASTM). US.

ASTM International. 2012. *Form and Style of Standards, ASTM Blue Book.* The American Society for Testing and Materials (ASTM). US.

Austin, E.W. 2015. *Strategic Public Relations Management.* 3rd Ed. Routledge. Oxfordshire, UK.

Averill, D. 2011. *Lean Sustainability: Creating Safe, Enduring, and Profitable Operations.* 1st Ed. Productivity Press. New York, US.

Ayers, J.C. 2017. *Sustainability: An Environmental Science Perspective.* 1st Ed. CRC Press. Florida, US.

Ayres, R.U.; van den Bergh, J.C.J.M.; Gowdy, J.M. 1998. *Viewpoint: Weak versus Strong Sustainability;* Tinbergen Institute Discussion Papers; Tinbergen Institute: Amsterdam, The Netherland/ pp. 98–103.

Azad, A.K. 2020. *Advances in Clean Energy Technologies.* 1st Ed. Academic Press. Cambridge, Massachusetts, US.

Baak, P.E. 2000. *Plantation Production and Political Power: Plantation Development in Southwest India in a Long Term Historical Perspective* 1743-1963. Oxford University Press. Oxford, UK.

Balaam, D.N., Dillman, B. 2018. *Introduction to International Political Economy.* 7th Ed. Routledge. Oxfordshire, UK.

Bali, P.S. 2011. *Quantity Food Production Operations and Indian Cuisine.* 1st Ed. Oxford University Press. Oxford, UK.

Barkley, A., Barkley, P.W. 2020. *Principles of Agricultural Economics.* 3rd Ed. Routledge. Oxfordshire, UK.

Barros-Velazquez, J. 2016. *Antimicrobial Food Packaging.* 1st Ed. Elsevier. Amsterdam, The Netherlands.

Barthwal, R.R. 2010. *Industrial Economics: An Introductory Textbook.* New Age International Publisher. Delhi, India.

Bartold, V.V., Suhrawardy, S. 2010. *Mussulman Culture.* 1st Ed. Oxford University Press. Oxford, UK.

Barut, Z.B., Ertekin, C., Karaagac, H.A. 2011. Tillage effects on energy use for corn silage in Mediterranean Coastal of Turkey. *Energy.* 36(9): 5466-5475.

Baugh, G. 2011. *The Fashion Designer's Textile Directory: A Guide to Fabrics' Properties, Characteristics, and Garment-Design Potential Paperback.* 1st ed. B.E.S. Publishing. New York, US.

Baxter, M. 1995. *Product Design.* 1th Ed. CRC Press. Florida, US.

Backlund, A. 2000. *The definition of system.* 1st Ed. In: Kybernetes.

Becerra-Fernandez, I., Sabherwal, R. 2014. *Knowledge Management Systems and Processes.* 2nd Ed. Routledge. Oxfordshire, UK.

Becker, G.S. Murphy, K.M. 2003. *Social Economics: Market Behavior in a Social Environment.* Belknap Press. London, UK.

Beeby, A.W., Narayanan, R.S. 2000. *Introduction to Design for Civil Engineers.* 1st Ed. CRC Press. Florida, US.

Bellah, B. 2015. *Sales Management.* 1st Ed. For Dummies. New Jersey, US.

Ben-Daya, M., Kumar, U., Murthy, D.N.P. 2016. *Introduction to Maintenance Engineering: Modelling, Optimization and Management.* 1st Ed. Wiley.

Benavides-Serrano, A.F., Pena-Sabogal, A.S., Leon, O.M., Sanchez Acevedo, G.H., Gonzalez Estrada, O.A. 2019. Optimization of parameters in material selection of tricone drill bit head design. *Journal of Physics Conference Series.* 1159(1): 012018.

Benna, U.G. Benna, A.U. 2018. *Crowdfunding and Sustainable Urban Development in Emerging Economies* (Advances in E-Business Research). 1st Ed. IGI Global. Pennsylvania, US.

Benson, M.H., Craig, R.K. 2017. *The End of Sustainability: Resilience and the Future of Environmental Governance in the Anthropogenic* (Environment and Society). University Press of Kansas. Kansas, US.

Berk, J. 2010. *Cost Reduction and Optimization for Manufacturing and Industrial Companies.* 1st Ed. Scrivener Publishing LLC. Massachusetts, US.

Berkel, N., Goncalves, J., Wac, K., Hosio, S., Cox, A.L. 2020. Human accuracy in mobile data collection. *International Journal of Human-Computer Studies.* 137, 102396.

Berlin, C., Adams, C. 2017. *Production Ergonomics: Designing Work Systems to Support Optimal Human Performance.* Ubiquity Press. London, UK.

Bernie, M. 1997. *Handbook of Electronics Manufacturing Engineering.* 1st Ed. Springer US.

Bernacki, H., Haman, J., Kanafojski, C.Z. 1972. Agricultural Machines, Theory and Construction, Volume 1. *Science Publication Foreign Cooperation Center of the Central Institute for Science, Technical and Economic Information,* Warsaw, Poland.

Best, K. 2006. *Design Management: Managing Design Strategy, Process and Implementation.* 1st Ed. AVA Publishing. West Sussex, UK.

Bharuchs, E. 2004. *Environmental Studies for undergraduate courses of all Branches of higher Education.* 6th Ed. University Grants Commission. Delhi, India.

Biazzo, S., Filippini, R. 2021. *Product Innovation Management: Intelligence, Discovery, Development.* Springer. Cham, Switzerland.

References

Bimber, B. 2000. The Study of Information Technology and Civic Engagement. *Political Communication*, 17(4): 329-333.

Bjelland, M., Kaplan, D., Malinowski, J., Getis, A. 2017. *Introduction to Geography.* 15th Ed. McGraw-Hill Education. New York City, US.

Blackwell, A.H., Manar, E. 2015. *Prototype.* 3rd Ed. UXL Encyclopedia of Science.

Blackwell, G.R. 2000. *The Electronic Packaging Handbook.* 1st Ed. CRC Press. Boca Raton. Florida, US.

Blake, G., Bly, R.W. 1993. *The Elements of Technical Writing.* Macmillan Publishers. New York, US.

Blakemore, K. 2013. *Social Policy: An Introduction.* 4th Ed. Open University Press. Maidenhead, Berkshire, UK.

Blank, L., Tarquin, A. 2017. *Engineering Economy.* 8th Ed. McGraw-Hill Education. New York, US.

Bolton, W. 1988. *Production Technology: Processes, Materials and Planning.* 1st Ed. Butterworth-Heinemann. Massachusetts, US.

Bourdaghs, M.K. 2014. *The Structure of World History: From Modes of Production to Modes of Exchange.* Translated. Karatani, K. Duke University Press. Durham, North Carolina, Us.

Braham, A. 2017. *Fundamentals of Sustainability in Civil Engineering.* 1st Ed. CRC Press. Florida, US.

Branden, N. 1997. *Taking Responsibility: Self-Reliance and the Accountable Life.* 1st Ed. Touchstone. New York, US.

Bratton, J., Gold, J. 2007. *Human Resource Management: Theory and Practice.* 4th Ed. Palgrave Macmillan. London, UK.

Brauner, P., Ziefle, M. 2015. Human Factors in Production Systems: Motives, Methods and Beyond. In: *Advances in Production Technology.* 1st Ed. *Cham.* Switzerland.

Brecher, C. 2015. *Advances in Production Technology.* 1st Ed. Springer. *Cham.* Switzerland.

Bridger, R.S. 2003. *Introduction to Ergonomics.* 1st Ed. Routledge. Oxfordshire, UK.

Campbell, D. 2007. *International Product Liability.* 2nd Ed. Lulu.com. Morrisville, North Carolina, US.

Brigham, E.F., Ehrhardt, M.C. 2016. *Financial Management: Theory & Practice.* 15th Ed. Cengage Learning. Massachusetts, US.

Brinkmann, R., Garren, S.J. 2018. 1st Ed. *The Palgrave Handbook of Sustainability: Case Studies and Practical Solutions.* Springer. Berlin, Germany.

Brock, D.A. 2016. *Sales Manager Survival Guide: Lessons from Sales' Front Lines.* 1st Ed. Partners in Excellence. Minnesota, US.

Brodner P. 1986. Skill Based Automated Manufacturing. 1st Ed. *Proceedings of the IFAC Workshop,* Karlsruhe, Federal Republic of Germany, 3-5 September.

Brown, C.V. 1999. IS Management Handbook. 7th Ed. Auerbach Publications. Boca Raton, Florida, US.

Brown, J.T. 2014. *The Handbook of Program Management: How to Facilitate Project Success with Optimal Program Management.* 2nd Ed. McGraw-Hill Education. New York, US.

Bryant, R., Knight, D.M. 2019. *The Anthropology of the Future.* Cambridge University Press. Cambridge, US.

Brundage, M.P., Chang, Q., Arinez, J., Xiao, G. 2015. *Reducing Costs in the Manufacturing Industry: An Energy Economic Perspective.* June 8-12, Charlotte, North Carolina, US.

Brunold, A. 2015. Civic Education for Sustainable Development and Its Consequences for German Civic Education Didactics and Curricula of Higher Education. *Discourse and Communication for Sustainable Education,* 6(1): 30-49.

Bryson, J.R., Sundbo, J., Fuglsang, L., Daniels, Peter. 2020. *Service Management: Theory and Practice.* Springer. Cham, Switzerland.

Budynas, R.G, Nisbett, J.K. 2014. *Shigley's Mechanical Engineering Design.* 10th Ed. McGraw-Hill Education. New York, US.

Bukhman, I. 2021. *Technology for Innovation: How to Create New Systems, Develop Existing Systems and Solve Related Problems.* Springer. Sims Cl, Singapore.

Burt, D., Petcavage, S., Pinkerton, R. 2011. *Proactive Purchasing in the Supply Chain: The Key to World-Class Procurement.* 1st Ed. McGraw-Hill. New York, Us.

Burton, S.J. 1995. *An Introduction to Law and Legal Reasoning.* 2nd Ed. Little Brown & Co Law & Business. New York City, US.

Bush, S.R., Oosterveer, P. 2020. *Governing Sustainable Seafood* (Earthscan Food and Agriculture). 1st Ed. Routledge. Oxfordshire, UK.

Byrd, R., DeMates, L. 2019. *Sustainability Made Simple: Small Changes for Big Impact.* Rowman & Littlefield Publishers. Washington, US.

Cairncross, A., Sinclair, P. 2014. *Introduction to Economics.* 6th Ed. Butterworth-Heinemann. Oxford, UK.

Campbell, F. 2021. *Human Factors: The Impact on Industry and the Environment. In: Natural Resources Management and Biological Sciences.* 1st Ed. IntechOpen. London, UK.

Caradonna, J.L. 2014. *Sustainability: A History.* 1st Ed. Oxford University Press. Oxford, UK.

Caradonna, J.L. 2018. *Routledge Handbook of the History of Sustainability.* 1st Ed. Routledge. Oxfordshire, UK.

Caradonna, J.L., Ballerini, E. 2016. *Sustainability. Brilliance Audio.* 1at Ed. Michigan, US.

Carayannis, E.G., Pirzadeh, A. 2021. *Culture, Innovation, and Growth Dynamics: A New Theory for the Applicability of Ideas.* Springer Palgrave Macmillan. London, UK.

Carbo, J.A., Dao, V.T., Haase, S.J., Blake Hargrove, M., Langella, I.M. 2017. *Social Sustainability for Business.* 1st Ed. Routledge. Oxfordshire, UK.

Cassandra, D., Josephson, M.D. 2009. Volume-reduced Products. Chapter 41. *Transfusion Medicine and Hemostasis: Clinical and Laboratory Aspects.* Elsevier. Amsterdam, The Netherlands.

Cato, M.S. 2011. *Environment and Economy.* 1st Ed. Routledge. Oxfordshire, UK.

Cavazzuti, M. 2013. *Optimization methods: from theory to design.* 1st Ed. Springer. Heidelberg, Germany.

Ceschin, F., Gaziulusoy, I. 2019. *Design for Sustainability: A Multi-level Framework from Products to Socio-technical Systems* (Routledge Focus on Environment and Sustainability). 1st Ed. Routledge. Oxfordshire, UK.

References

Cezarina, A., Frumuşanu, G. 2017. A Review on Optimization of Manufacturing Process Performance. *International Journal of Modeling and Optimization* 7(3):139-144.

Chen, Z., Li, P., Jiang, S., Chen, H., Wang, J. Cao, C. 2021. Evaluation of resource and energy utilization, environmental and economic benefits of rice water-saving irrigation technologies in a rice-wheat rotation system. *Science of The Total Environment.* 757: 143748.

Chase, H. 2020. *Handbook on designing for quantity production.* 2nd Ed. Mcgrwaw-hill, London, UK.

Chaudhry, S., Bansal, A., Krishan, G. 2014. Finite element analysis and weight reduction of universal joint using CAE tools. *International Journal of Engineering Research & Technology.* 3(10): 1137-1139.

Cherchye, L., Kuosmanen, T., Leleu, H. 2010. Technical and Economic Efficiency Measures Under Short Run Profit Maximizing Behavior. *Recherches économiques de Louvain.* 76(2):163-173.

Childs, P.R.N. 2014. *Mechanical Design Engineering Handbook.* 3rd Ed. Elsevier, Amsterdam, The Netherlands.

Childs, P.R.N. 2013. *Mechanical Design, Engineering Handbook.* 2nd Ed. Elsevier Butterworth-Heinemann. Oxford, UK.

Childs, T.H.C. 2021. *Mechanical Design, Theory and Applications.* 3rd Ed. Butterworth-Heinemann. Vermont, US.

Ching, F.D.K., Eckler, J.F. 2012. *Introduction to* Architecture. 1st Edition. 1st Ed. Wiley. Ontario, Canada.

Choi, Y.K. 2004. *Principles of Applied Civil Engineering Design.* 1st Ed. ASCE Press. Virginia, US.

Christensen, N. 2012. *The Environment and You.* 1st Ed. Benjamin Cummings. San Francisco, California, US.

Chryssolouris, E.L.K.E. 2006. *Manufacturing Systems: Theory and Practice.* Springer-Verlag. New York, US.

Ciulla, J.B., Martin, C. Solomon, R.C. 2013. *Honest Work: A Business Ethics Reader.* 3rd Ed. Oxford University Press. Oxford, UK.

Cimorelli, S. 2016. *Kanban for the Supply Chain: Fundamental Practices for Manufacturing Management.* 2nd Ed. Productivity Press. Oxfordshire, UK.

Clarke, R.G. 1991. *Industrial Economics.* 1st Ed. Wiley-Blackwell. New Jersey, US.

Claus, P., Marriott, J. 2017. *History: An Introduction to Theory, Method and Practice.* 2nd Ed. Routledge. Oxfordshire, UK.

Clayton, T., Radcliffe, N. 2016. *Sustainability: A Systems Approach.* 1st Ed. Routledge. Oxfordshire, UK.

Clifford, M.J., Perrons, R.K., Ali, S.H., Grice, T.A. 2020. *Extracting Innovations, Mining, Energy, and Technological Change in the Digital Age.* 1st Ed. CRC Press. Florida, US.

Coady, T., May, E., Figueres, C. 2020. *Rebuilding Earth: Designing Ecoconscious Habitats for Humans.* 1st Ed. North Atlantic Books California, US.

Cocklin, C. 1995. Agriculture, society and environment: discourses on sustainability. *International Journal of Sustainable Development & World Ecology.* 2(4): 240-256.

Coffelt, D., Hendrickson, C. 2017. *Fundamentals of Infrastructure Management.* 1st Ed. Pittsburgh. Pennsylvania, US.

Cohen, S. 2014. *Sustainability Management: Lessons from and for New York City, America, and the Planet.* 1st Ed. Columbia University Press. New York, US.

Colomer, J.M. 2010. *The Science of Politics: An Introduction.* 1st Ed. Oxford University Press. Oxford, UK.

Colvin, E.M. 2017. Transportation of Agricultural Products in the United States, 1920-June 1939, Vol. 2: *A Selected List of References Relating to the Various Phases of Highway, Rail, and Water Transportation.* 1st Ed. Forgotten Books. London, UK.

Cooper, T. 2010. The significance of product longevity. In Cooper, T. *Longer Lasting Products: alternatives to the throwaway society.* Routledge. Oxfordshire, UK.

Cooper, T. 2010. The significance of product longevity. In Cooper, T. *Longer Lasting Products: alternatives to the throwaway society.* Farnham: Gower. 3-36.

Cooper, T. 1994. *Beyond Recycling: the longer life option.* New Economics Foundation. London, UK.

Cooper, R. 1984. The Performance Impact of Product Innovation Strategies. *European Journal of Marketing.* 18(5): 5-54.

Cooper, R., Kleinschmidt, E.J. 1995. New Product Performance: Keys to Success, Profitability and Cycle Time Reduction. *Journal of Marketing Management.* 11(4):315-337.

Coovert, M.D., Thompson, L.F. 2014. *The Psychology of Workplace Technology.* 1st Ed. Routledge. Oxfordshire, UK.

Connolly, W. 1981. *Appearance and Reality in Politics.* Cambridge University Press. Cambridge, UK.

CORE. 2019. *The Economy: Economics for a Changing World.* 1st Ed. Oxford University Press. Oxford, UK.

Coronel, C.M., Morris, S. 2018. *Database Systems: Design, Implementation, & Management.* 13th Ed. Cengage Learning. Massachusetts, US.

Corradini, M.G. 2018. Shelf Life of Food Products: From Open Labeling to Real-Time Measurements. *Annual Review of Food Science and Technology* 9(1): 251-269.

Courage, C., Baxter, K. 2005. *Understanding Your Users: A Practical Guide to User Requirements Methods, Tools, and Techniques.* Morgan Kaufmann Publishers.

Cox, J., Griffith, S., Giorgi, S., King, G. 2013. Consumer understanding of product lifetimes. Resources, *Conservation & Recycling.* 79: 21–29.

Cramer, G., Paudel, K., Schmitz, A. 2019. *The Routledge Handbook of Agricultural Economics.* Routledge. Oxfordshire, UK.

Crawford, C.M., Di Benedetto, C.A. 2014. *New Products Management.* 11th Ed. McGraw-Hill Education. New York, US.

Dadd-Redalia, D. 1994. *Sustaining the earth: choosing consumer products that are safe for you, your family, and the earth.* 1st Ed. Hearst Books. New York, US.

Daim, T.U., Kim, J., Phan, K. 2017. *Research and Development Management: Technology Journey through Analysis, Forecasting and Decision Making.* 1st Ed. Springer. Cham, Switzerland.

Dalkir, K. 2005. *Knowledge Management in Theory and Practice.* 1st Ed. Butterworth-Heinemann. Massachusetts, US.

Daniel, D.E. 1993. *Geotechnical Practice for Waste Disposal.* Springer. New York, US.

Davies, S. 1991. *Definitions of Art.* Cornell University Press. New York, US.

References

Davim, J.P. 2012. *Statistical and Computational Techniques in Manufacturing.* 1st Ed. Springer Nature. Aktiengesellschaft, Switzerland.

Dawson, P., Andriopoulos, C. 2021. *Managing Change, Creativity and Innovation.* 4th Ed. SAGE Publications. New York, US.

de Jong, E. 2014. Personal Development for Beginners: Book 1 - 3: *Goal Setting for Success; Time Management for a Productive Life; The Power of Habit: be Efficient in Everything you do.* CreateSpace Independent Publishing Platform. California, US.

Desjardins, R., Schuller, T. 2006. *Measuring the Effects of Education on Health and Civic Engagement: Proceedings of the Copenhagen Symposium.* OECD. Paris, French.

Devine, P.J., Lee, N., Jones, R.M., Tyson, W.T. 1985. *An Introduction to Industrial Economics.* 1st Ed. Routledge. Oxfordshire, UK.

Dietrich, D.M., Kenworthy, M., Cudney, E.A. 2019. *Additive Manufacturing Change Management: Best Practices.* 1st Ed. CRC Press. Florida, US.

Dixit, A. 2014. *Microeconomics: A Very Short Introduction* (Very Short Introductions). Oxford University Press. Oxford, UK.

Dobson, S. 2004. *Introduction to Economics.* Oxford University Press. Oxford, UK.

Dodds, F., Duque Chopitea, C., Ruffins, R. 2021. *Tomorrow's People and New Technology: Changing How We Live Our Lives.* 1st Ed. Routledge. Oxfordshire, UK.

Doppelt, B. 2009. *Leading Change toward Sustainability.* 2nd Ed. Routledge. Oxfordshire, UK.

Dransfield, R. 2013. *Business Economics.* 1st Ed. Routledge. Oxfordshire, UK.

Drummond, H.E., Goodwin, J.W. 2010. *Agricultural Economics.* 3rd Ed. Pearson. New York, US.

DuBrin, A.J. 2011. *Essentials of management.* 9th Ed. South-Western College Pub. Ohio, US.

Dul, J., Weerdmeester, B.A. 2008. Ergonomics for Beginners. 3rd Ed. Taylor & Francis. Oxfordshire, UK.Daynard, R.A., Legresley, E. 2012. Product liability. *Tobacco Control.* 21(2): 227-8.

Duncan, C., Richardson, P.E. 2014. *Plant Equipment and Maintenance Engineering Handbook.* McGraw-Hill Education. New York City, US.

Durmaz, Y., Yasar, H.V. 2016. Brand and Brand Strategies. International Business Research 9(5): 48-56.

Dye, T.R. 1976. *Policy Analysis.* University of Alabama Press. Tuscaloosa, Alabama, US.

Eccleston, C.H. 2011. *Environmental Impact Assessment: A Guide to Best Professional Practices.* 1st Ed. CRC Press. New Jersey, US.

Ee, S. 2018. *What is Facilities Management All About?: The practice of facilities management for today's dynamic business environment.* CreateSpace Independent Publishing Platform. California, US.

Efraim, T. Carol, P. 2011. *Information Technology Management.* 8th Ed. Wiley. New Jersey, US.

Eide, A. Jenison, R. Mashaw, L. Northup, L. 2002. *Engineering: Fundamentals and Problem Solving.* McGraw-Hill Companies. New York, US.

Eissen, K., Steur, R. 2014. *Sketching, Product Design Presentation.* Laurence King Publishing. London, UK.

References

Elkington, J. 1994. Towards the sustainable corporation: Win-win business strategies for sustainable development. *Calif. Manage. Rev.* 36, 90–100.

Emerson, T., Stewart, M. 2011. *The Learning and Development Book.* Association for Talent Development. Virginia, US.

Emmitt, S. 2017. *Design Management.* 1st Ed. Routledge. Oxfordshire, UK.

Embley, D.W., Thalheim, B. 2011. *Handbook of Conceptual Modeling: Theory, Practice, and Research Challenges.* Springer-Verlag, Berlin, Germany.

Emblem, A., Emblem, H. 2012. *Packaging Technology: Fundamentals, Materials and Processes.* 1st Ed. Woodhead Publishing. Sawston, UK.

Epstein, E. 2015. *Disposal and Management of Solid Waste.* 1st Ed. CRC Press. New Jersey, US.

Evans, S., Cooper, T. 2010. Consumer Influences on Product Life-Spans. In Cooper, T. *Longer Lasting Products.* Routledge. Oxfordshire, UK.

Falade, B.A., Coultas, C.J. 2017. Scientific and non-scientific information in the uptake of health information: The case of Ebola. *South African Journal of Science.* 113: 1-8.

Fanchi, J.R. 2005. *Principles of Applied Reservoir Simulation.* 3rd Ed. Gulf Professional Publishing. London, UK.

Farraji, H., Qamaruz Zaman, N., Mohajeri, P. 2016. *Waste Disposal: Sustainable Waste Treatments and Facility Siting Concerns.* 1st Ed. Chapter 3. Aziz, H., Amr, S. Control and Treatment of Landfill Leachate for Sanitary Waste Disposal. IGI global disseminator of knowledge. Pennsylvania, US.

Fast, L.E. 2015. *The 12 Principles of Manufacturing Excellence: A Lean Leader's Guide to Achieving and Sustaining Excellence.* 1st Edition. Productivity Press. Oxfordshire, UK.

Farmer, N. 2013. *Trends in Packaging of Food, Beverages and Other Fast-Moving Consumer Goods* (FMCG). 1st Ed. Woodhead Publishing. Sawston, UK.

Farokhzad, S., Modares Motlagh, A., Ahmadi Moghadam, P., Jalali Honarmand, S., Kheiralipour, K. 2020. Application of infrared thermal imaging technique and discriminant analysis methods for non-destructive identification of fungal infection of potato tubers. *Journal of Food Measurement and Characterization* 14 (1), 88-94.

Fayos-Solà, E., Cooper, C. 2018. *The Future of Tourism: Innovation and Sustainability.* 1st Ed. Springer. Berlin, Germany.

Fenwick, T. 2016. *Professional Responsibility and Professionalism: A sociomaterial examination.* 1st Ed. Routledge. Oxfordshire, Uk.

Fernald, L.D. 2008. *Psychology: Six perspectives.* Thousand Oaks, CA: Sage Publications.

Fischer, F., Miller, G.J., Sidney, M.S. 2007. *Handbook of Public Policy Analysis Theory, Politics, and Methods.* CRC Press.

Fisher, A.G. B. 1939. Production: primary, secondary and tertiary. *Economic Record.* 15(1), 24-38.

Fisher, I. 2006. *Elementary Principles of Economics.* 1st Ed. Cosimo Classics. New York, US.

Forbes J. D. 2016. *The Consumer Interest (RLE Consumer Behaviour) Dimensions and Policy Implications.* 1st Ed. Routledge. Oxfordshire, UK.

Francis, L. 2015. *Materials Processing.* 1st Ed. Academic Press. Massachusetts, US.

Freedman, L. 2013. *Strategy.* Oxford University Press. ISBN 978-0-19-932515-3.

References

Frick, T. 2016. *Designing for Sustainability: A Guide to Building Greener Digital Products and Services*. 1st Ed. O'Reilly Media. California, US.

Fricker, J.D., Whitford, R.K. 2004. *Fundamentals of Transportation Engineering: A Multimodal Systems Approach*. 1st Ed. Prentice Hall. Hoboken, New Jersey, US.

Friedman, J. 2020. *Managing Sustainability: First Steps to First Class*. Business Expert Press. New Jersey, US.

Fucks, R., Harland, R., Giddens, A. 2015. *Green Growth, Smart Growth: A New Approach to Economics, Innovation and the Environment (Anthem Environment and Sustainability Initiative* (AESI)). 1st Ed. Anthem Press. London, US.

Fukuyama, F. 2008. State building in the Solomon Islands. *Pac. Econ. Bull.* 23, 1–17.

Furrer, O. 2006. *Marketing Strategies. In: Marketing Management: International Perspectives*. 1st Ed. Vijay Nicole Publishing. Chennai, India.

Galanakis, C. 2019. *Food Quality and Shelf Life*. 1st Edition. Academic Press. Cambridge, Massachusetts, US.

Gamma, E., Helm, R., Johnson, R., Vlissides, J. 1995. *Design Patterns*. Addison-Wesley Professional.

Garcia, R. 2014. *Creating and Marketing New Products and Services*. 1st Ed. Auerbach Publications. Florida, USA.

Garcia-Arca, J., González-Portela Garrido, A.T., Prado-Prado, J.C. 2017. Sustainable Packaging Logistics: The link between ustainability and Competitiveness in Supply Chains. *Sustainability*. 9: 1098.

Gardoni, P. 2018. *Routledge Handbook of Sustainable and Resilient Infrastructure*. 1st Ed. Routledge. Oxfordshire, UK.

Gates, B. 2021. *How to Avoid a Climate Disaster: The Solutions We Have and the Breakthroughs We Need*. Knopf. New York, US.

Gaynor, G.H. 1996. *Handbook of Technology Management*. 1st Ed. McGraw-Hill. New York, US.

Geographic, N. 2009. *The Knowledge Book: Everything You Need to Know to Get by in the 21st Century*. National Geographic. Iowa, US.

Geographic, N. 2011. *The Science Book: Everything You Need to Know About the World and How It Works*. National Geographic. Washington, US.

Gerald, J., Friedman, D.R. 2010. *Wholesale Packing Resource Guice*. New Entry Sustainable Farming Project. Boston. Massachusetts, US.

Gibson, P., Greenhalgh, G., Kerr, R. 1995. *Manufacturing Management: Principles and Concepts*. Springer. Berlin, Germany.

Gilbert, P., Bobadilla, N., Gastaldi, L., Le Boulaire, M., Lelebina, O. 2018. 1st Ed. *Innovation, Research and Development Management*. Wiley. New Jersey, US.

Ghodeswar, B.M. 2008. Building brand identity in competitive markets: A conceptual model. *Journal of Product & Brand Management*. 17: 4-12.

Ghorbani, R., Mondani, F., Amirmoradi, S., Feizi, H., Khorramdel, S., Teimouri, M., Sanjani, S., Anvarkhah, S., Aghel, H. 2011. A case study of energy use and economical analysis of irrigated and dryland wheat production systems. *Applied Energy*. 88(1): 283-288.

Ghodeswar, B.M. 2008. Building brand identity in competitive markets: A conceptual model. *Journal of Product & Brand Management*. 17: 4-12.

References

Ghosh, N. 2019. *Livestock Production Management.* 1st Ed. PHI Learning. Delhi, India.

Giddens, A., Duneier, M., Applebaum, R. 2007. *Introduction to Sociology.* Sixth Edition. New York: W.W. Norton and Company.

Gift, S.J.G., Maundy, B. 2021. *Electronic Circuit Design and Application.* 1st Ed. Springer. Berlin, Germany.

Gitlow, H.S. 2000. *Quality Management Systems: A Practical Guide.* CRC Press. New Jersey, US.

Glass, M.R., Rose-Redwood, R. 2015. *Performativity, Politics, and the Production of Social Space.* 1st Ed. Routledge. Oxfordshire, UK.

Goldblatt, D.L. 2005. *Sustainable Energy Consumption and Society: Personal, Technological, or Social Change?* Springer. Berlin, Germany.

Goodship, V. 2010. *Management, recycling and reuse of waste composites.* Woodhead Publishing. Oxford, UK.

Goodstein, E.S., Polasky, S. 2020. *Economics and the Environment.* 9th Ed. Wiley. Hoboken, New Jersey, US.

Gouge, M., Michaleris, P. 2018. *Thermo-Mechanical Modeling of Additive Manufacturing.* 1st Ed. Elsevier. Amsterdam, The Netherlands.

Graham, B., Zweig, J. 2006. The Intelligent Investor: The Definitive Book on Value Investing. *A Book of Practical Counsel.* Harper Business. New York, US.

Greene, J. 2011. *Plant Design, Facility Layout, Floor Planning.* 1st Ed. CreateSpace Independent Publishing Platform. Scotts Valley, California, US.

Gregory, P.R., Stuart, R.C. 2013. *The Global Economy and Its Economic Systems.* 1st Ed. Cengage Learning. Massachusetts, US.

Greiner, D. 1997. *The Basics of Idea Generation.* 1st Ed. Productivity Press. New York, US.

Griffiths, H., Strayer, E., Cody-Rydzewski, S. 2019. *Introduction to Sociology.* 2nd Ed. XanEdu Publishing. Ann Arbor, Michigan, US.

Grober, U., Cunningham, R. 2012. *Sustainability: A Cultural History.* 1st Ed. Green Book. Cambridge, UK.

Grohens, Y., Kumar, S.K., Boudenne, A., Weimin, Y. 2021. *Recycling and Reuse of Materials and Their Products.* 1st Ed. Apple Academic Press. New York, US.

Grohens, Y., Sadasivuni, K.K., Boudenne, A. 2013. *Recycling and Reuse of Materials and Their Products. Advances in Materials Science.* Volume 3. 1st Ed. Apple Academic Press Inc. Ontario, Canada.

Groover, M.P. 2011. *Introduction to Manufacturing Processes.* 1st Ed. Wiley. Hoboken, New Jersey, US.

Gryna, F., Chua, R., Defeo, J. 2005. *Juran's Quality Planning and Analysis for Enterprise Quality.* 5th Ed. McGraw-Hill. New York, US.

Guinee, J.B. 2002. *Handbook on life cycle assessment operational guide to the ISO standards.* Kluwer Academic, New York, US.

Gupta, A.K. 2014. *Engineering management.* 1st Ed. S. Chand Publishing. Delhi, Indian.

Gupta, M. 2019. *Installation Maintenance and Repair of Electrical Machines and Equipments.* 2nd Ed. S.K. Kataria & Sons. New Delhi, Delhi, India.

Gupta, R.K., Gupta, H. 2019. *Working Capital Management & Finance: A Hand Book for Bankers and Finance Managers.* Notion Press. Chennai, India.

References

Gupta, S.K. 2020. *Horticultural Crop Production.* 1st Ed. Rajat Publication. New Delhi, Indi.

Gupta, S., Starr, M. 2014. *Production and Operations Management Systems.* 1st Ed. CRC Press.

Gurjar, B.R., Molina, L.T., Ojha, C.S.P. 2010. *Air Pollution: Health and Environmental Impacts.* 1st Ed. CRC Press. New Jersey, US.

Gurr, A. 2018. *The Effects of Positive and Negative Environmental Responsibility on Financial Performance.* CMC Senior Theses. 1800.

Haefner, J.W. 2005. *Modeling Biological Systems: Principles and Applications.* 1st Ed. Springer. New York, US.

Hage, J., Akkermans, B. 2014. *Introduction to Law.* Springer. Berlin, Germany.

Haksever, C., Render, B. 2013. *Service Management: An Integrated Approach to Supply Chain Management and Operations.* 1st Ed. Pearson. New York, US.

Hammes, A. 2015. *Stress-Free Sustainability: Leverage Your Emotions, Avoid Burnout and Influence Anyone.* 2nd Ed. CreateSpace Independent Publishing Platform. California, US.

Hanley, N., Shogren, J., White, B. 2019. *Introduction to Environmental Economics.* 3rd Ed. Oxford University Press. Oxford, UK.

Harland, C.M. (1996) *Supply Chain Management, Purchasing and Supply Management,* Logistics, Vertical.

Harrington, H.J. 2018. *Creativity, Innovation, and Entrepreneurship: The Only Way to Renew Your Organization.* 1st Ed. Productivity Press. New York, US.

Harris, P. 2016. *An Introduction to Law.* 8th Ed. Cambridge University Press.

Harris, D. Harris, S. 2012. *Digital Design and Computer Architecture.* 2nd Ed. Morgan Kaufmann. California, US.

Harris, N. 2001. *Business Economics: Theory and Application.* 1st Ed. Routledge. Oxfordshire, UK.

Harris, R., Harris, C., Wilson, E., Womack, J., Jones, D., Shook, J., Ferro, J. 2003. *Making Materials Flow: A Lean Material-Handling Guide for Operations, Production-Control, and Engineering Professionals.* 1st Ed. Lean Enterprises Inst. Massachusetts, US.

Harrison, R.M., Hester, R.E. 2012. *Environmental Impacts of Modern Agriculture.* 1st Ed. Royal Society of Chemistry. Burlington House; London, UK.

Hartman, H.L. 2021. *Introductory Mining Engineering.* 2nd Ed. Wiley. New Jersey, US.

Harwood, S.A. 2013. *ERP: The Implementation Cycle.* 1st Ed. Routledge. Oxfordshire, UK.

Hauck, J. 2009. *Electrical Design of Commercial and Industrial Buildings.* 1st Ed. Jones & Bartlett Learning. Massachusetts, US.

Hayes, B. 2016. *Infrastructure: A Guide to the Industrial Landscape.* 1st Ed. W. W. Norton & Company. New York, US.

Hayes, B. 2006. *Infrastructure: the book of everything for the industrial landscape.* 1st Ed. W.W. Norton & Company. New York, US.

Heidari, M. D., Omid, M. & Akram, A. (2011). Energy efficiency and econometric analysis of boiler production farms. *Energy,* 36(11), 6536-41.

Heidorn, K., Whitelaw, I. 2020. *The Field Guide to Natural Phenomena: The Secret World of Optical, Atmospheric and Celestial Wonders.* 1st Ed. Firefly Books. Richmond Hill, ON, Canada.

Heiskanen, E., 1996. Conditions for product life extension. *Natl. Consum. Res. Cent. Work. Pap. 22*, 1996.

Hellemans, B. 2017. *Understanding Culture, A Handbook for Students in the Humanities.* 1st Ed. Amsterdam University Press. Amsterdam, The Netherlands.

Hellier, C. 2012. *Handbook of Nondestructive Evaluation.* 2nd Ed. McGraw-Hill Education. New York City, US.

Herring H., Sorrell, S. 2008. *Energy Efficiency and Sustainable Consumption: The Rebound Effect* (Energy, Climate and the Environment). 1st Ed. Palgrave Macmillan. London, UK.

Hill, M., Varone, F. 2021. *The Public Policy Process.* 8th Ed. Routledge. Oxfordshire, UK.

Hylland Eriksen, T. 2004. *What is Anthropology?* Pluto. London.

Hodges, C., Sekula, M. 2013. *Sustainable Facility Management - The Facility Manager's Guide to Optimizing Building Performance.* 1st Ed. CreateSpace Independent Publishing Platform. California, US.

Holechek, J.L., Cole, R.A., Fisher, J.T., Valdez, R. 2002. *Natural Resources: Ecology, Economics, and Policy.* 2nd Ed. Pearson. New York City, New York, US.

Hollins, B., Pugh, S. 1990. *Successful Product Design: What to Do and When.* 1st Ed. Butterworth-Heinemann. London, UK.

Holmgren, D. 2002. *Permaculture: Principles and Pathways beyond Sustainability.* 1st Ed. Holmgren Design Services. Fremantle, Australia.

Homburg, C., Schwemmle, M., Kuehnl, C. 2015. New Product Design: Concept, Measurement, and Consequences. *Journal of Marketing,* 79(3), 41-56.

Hotelling, H. 1931. The economics of exhaustible resources. *J. Polit. Econ.* 39, 137–175.

Howard, P.H. 2020. *Handbook of Environmental Degradation Rates.* 1st Ed. CRC Press. New Jersey, US.

Hudson, R. 2020. *Co-produced Economies: Capital, Collaboration, Competition.* 1st Ed. Routledge. Oxfordshire, UK.

Hui, Y.H. et al. 2007a. *Handbook of Food Products Manufacturing: Health, Meat, Milk, Poultry, Seafood, and Vegetables,* Volume 2. 1st Ed. John Wiley & Sons, Inc. New Jersey, USA.

Hui, Y.H. et al. 2007b. *Handbook of Food Products Manufacturing. Principles, Bakery, Beverages, Cereals, Cheese, Confectionary, Fats, Fruits, and Functional Foods.* 1st Ed. John Wiley & Sons, Inc. New Jersey, USA.

Hulse, K.L. 2000. *Anode manufacture: raw materials, formulation and processing parameters.* 1st Ed. R & D Carbon Limited. Granges, Switzerland.

Hunt, E. 2010. *Human Intelligence.* 1st Ed. Cambridge University Press. Cambridge, UK.

Hurmuzlu, Y., Nwokah, O.D.I. 2001. *The Mechanical Systems Design Handbook: Modeling, Measurement, and Control.* 1st Ed. CRC Press. Florida, US.

Hussen, A. 2019. *Principles of Environmental Economics and Sustainability.* 4th Ed. Routledge. Oxfordshire, UK.

Hutchin, T. 2002. *Constraint Management in Manufacturing Optimising the Supply Chain.* 1st Ed. CRC Press.

References

Hutt, M.D. Speh, T.W. 2012. *Business Marketing Management: B2B.* 11th Ed. South-Western College Pub. Ohio, US.

IEEE, 1990. *IEEE Standard Computer Dictionary: A Compilation of IEEE Standard Computer Glossaries.* Institute of Electrical and Electronics Engineers (IEEE). New York, US.

Ike, D.N. 1984. The System of Land Rights in Nigerian Agriculture. *Amer. J. Econ. Sociol.* 43, 469–480.

Ingersoll R., Kostof, S. 2012. World Architecture: A Cross-Cultural History. 1st Ed. Oxford University Press. Oxford, UK.

Ingenbleek, P., Frambach, R.T., Verhallen, T.M.M., 2013. Best Practices for New Product Pricing: Impact on Market Performance and Price Level under Different Conditions. *Journal of Product Innovation Management.* 30(3): 560-573.

Inglis, D. 2005. *Culture and Everyday Life.* 1st Ed. Routledge. Oxfordshire, UK.

Ingram, T.N., LaForge, R.W., Avila, R.A., Schwepker Jr. C.H., Williams, M.R. 2019. *Sales Management: Analysis and Decision Making.* 10th Ed. Routledge. Oxfordshire, UK.

ISO 14044. 2006. *Environmental management-life cycle assessment-principles and framework.* Geneva: International Organization for Standardization.

Ivanov, D., Tsipoulanidis, A., Schönberger, J. 2017. Production Strategy. Chapter 6. *Global Supply Chain and Operations Management: A decision-oriented introduction into the creation of value, Cham,* 1st Ed. Springer Nature. Berlin, Germany.

Jacques, P. 2014. *Sustainability: The Basics.* 1st Ed. Routledge. Oxfordshire, UK.

Jahanbakhshi, A., Heidari Raz Dareh, S., Kheiralipour, K. 2020. Finite Element Fatigue Analysis of Mouldboard Plough Cross Bar Based on the Draft Force of MF 399 Tractor. *Journal of Failure Analysis and Prevention.* 20(4): 2106-2110.

Jain, E.R.K. 2010. *Production Technology: Manufacturing Processes, Technology and Automation.* 17th Ed. Khanna Publishers. Delhi, India.

Jameel, F., Hershenson, S., Khan, M.A., Martin-Moe, S. 2015. *Quality by Design for Biopharmaceutical Drug Product Development.* Springer-Verlag. New York, US.

James, P. 2014. *Urban Sustainability in Theory and Practice.* Routledge. Oxfordshire, UK.

Jamshidian, M., Kamelnia, S., Varzgani, M., Rais Sadati, S.M. 2013. *Comprehensive Reference for Iranian and Islamic Fashion Design.* Qalam Azin Reza Publication. Mashhad, Iran.

Jaques, E., Cason, K. 1994. *Human Capability: A Study of Individual Potential and Its Application.* Cason Hall & Co Pub. Maryland, US.

Jedlicka, W. 2008. *Packaging Sustainability: Tools, Systems and Strategies for Innovative Package Design.* 1st Ed. Wiley. New Jersey, Us.

Jefferis, A. Madsen, D.A. Madsen, D.P. 2016. *Architectural Drafting and Design.* 7th Ed. Cengage Learning. Delhi, India.

Jeffus, L. 2011. *Welding and Metal Fabrication.* 1st Ed. Cengage Learning. Massachusetts, US.

Jensen, P.A., van der Voordt, T. 2016. *Facilities Management and Corporate Real Estate Management as Value Drivers: How to Manage and Measure Adding Value.* 1st Ed. Routledge. Oxfordshire, UK.

Jevons, W.S. 2008. *The Theory of Political Economy.* Kessinger Publishing. Whitefish, Montana, US.

Jiang, W., Vlahopoulos, N., Castanier, M.P., Thyagarajan, R., Mohammad, S. 2015. Tuning material and component properties to reduce weight and increase blast worthiness of a notional V-hull structure. *Case Studies in Mechanical Systems and Signal Processing.* 2: 19-28.

Joanna, M., Bassert, V.M.D. 2021. *Clinical Textbook for Veterinary Technicians and Nurses.* 10th Ed. Saunders, US.

Johansson, A. 1992. *Clean Technology.* 1st Ed. CRC Press. Boca Raton, Florida, US.

Jovanovic, M. 2012. *Evolutionary Economic Geography.* 1st Ed. Routledge. Oxfordshire, UK.

Jovanovic, M. 2001. *Geography of Production and Economic Integration.* 1st Ed. Routledge. Oxfordshire, UK.

Jurgens, U. 2000. *New Product Development and Production Networks.* 1st ed. Springer-Verlag Berlin Heidelberg, Germany.

Kadane, S.M. Bhatwadekar, S.G. 2011. Manufacturing Facility Layout Design and Optimization Using Simulation. *International Journal of Advanced Manufacturing Systems.* 2(1): 59-65.

Kahn, K.B. 2012. *The PDMA handbook of new product development.* 3rd Ed. John Wiley & Sons. New Jersey, US.

Kalpakjian, S., Schmid, S.R. 2013. *Manufacturing Engineering and Technology.* 7th Ed. Pearson Canada, Canada.

Karr, S., Houtman, A., InterlandI, J. 2021. *Environmental Science for a Changing World.* 4th Ed. W.H. Freeman & Co. Ltd. New York, US.

Kassel, K. 1750. *The Thinking Executive's Guide to Sustainability* (Environmental and Social Sustainability for Business Advantage Collection). 1st Ed. Business Expert Press. New Jersey, US.

Kaya, D., Canka Kilic, F., Ozturk, H.H. 2021. *Energy Management and Energy Efficiency in Industry: Practical Examples.* Springer. Cham, Switzerland.

Keinonen, T.K., Takala, R., 2006. *Product Concept Design: A Review of the Conceptual Design of Products in Industry.* 1st Ed. Springer-Verlag. London, UK.

Kelsen, H. 2005. *General Theory of Law and State.* 1st Ed. Routledge. Oxfordshire, UK.

Kenen, P.B. 2000. *The International Economy.* 4th Ed. Cambridge University Press. Cambridge, UK.

Khan, W.A., Abdul Raouf, S.I. 2005. *Standards for Engineering Design and Manufacturing.* 1st Ed. CRC Press. Boca Raton, Florida, US.

Kheiralipour, K. 2021a. Environmental Life Cycle Assessment of Poultry Production Systems. Chapter 3. Jacob-Lopes, E., Zepka, L. Q., Deprá, M. C. *Interdisciplinary applications of the life cycle assessment tool.* 1st Ed. Nova Science Publishers, New York.

Kheiralipour, K. 2021b. Management knowledge and technology in agriculture and natural resources. *13th Iranian National Congress of Biosystems Engineering and Agricultural Mechanization.* 15-17 September, Tehran, Iran. (In Persian).

Kheiralipour, K. 2020. *Environmental Life Cycle Assessment.* 1st Edition. Ilam University Publication. Ilam, Iran. (In Persian).

References

Kheiralipour, K., Ahmadi, H., Rajabipour, A., Rafiee, S. 2018. *Thermal Imaging, Principles, Methods and Applications.* 1st Ed. Ilam University Publication, Ilam, Iran. (In Persian).

Kheiralipour, K., Ahmadi, H., Rajabipour, A., Rafiee, S., Javan-Nikkhah, M. 2015. Classifying healthy and fungal infected-pistachio kernel by thermal imaging technology. *International Journal of Food Properties.* 18(1): 93-99.

Kheiralipour, K., Ahmadi, H., Rajabipour, A., Rafiee, S., Javan-Nikkhah, M., Jayas, D.S., Siliveru, K. 2016. Detection of fungal infection in pistachio kernel by long-wave near-infrared hyperspectral imaging technique. *Quality Assurance and Safety of Crops & Foods.* 8(1): 129-135.

Kheiralipour, K., Jafari Samrin, H., Soleimani, M. 2017a. Determining the environmental impacts of canola production by life cycle assessment, case study: Ardabil Province. *Iranian Journal of Biosystems Engineering,* 48(4), 517-526.

Kheiralipour, K., Payandeh, Z. Khoshnevisan, B. 2017b. Evaluation of Environmental Impacts in Turkey Production System in Iran. *Iranian Journal of Applied Animal Science,* 7, 507-512.

Kheiralipour, K., Sheikhi, N. 2020. Material and energy flow in different bread baking types. *Environment, Development and Sustainability.* 23 (7), 10512-10527.

Kheiralipour, K., Pormah, A. 2016. Introducing new shape features for classification of cucumber fruit based on image processing technique and artificial neural networks. *Journal of food process engineering.* 40(6): 12558.

Khojasteh, Y. 2018. *Production Management Advanced Models, Tools, and Applications for Pull Systems.* 1st Ed. Productivity Press. New York, US.

Khoobbakht, G., Karimi, M., Kheiralipour, K. 2019. Effects of biodiesel-ethanol-diesel blends on the performance indicators of a diesel engine: A study by response surface modeling. *Applied Thermal Engineering.* 148: 1385-1394.

Kimball, D.S. 1933. The Social Effects of Mass Production. Science. New Series, 77(1984): 1-7.

Kiran, D.R. 2018. *Production Planning and Control: A Comprehensive Approach.* 1st Ed. Butterworth-Heinemann, Elsevier, Oxford, United Kingdom.

Kiran, R.U. 2008. *Textbook of Technology Management.* Laxmi Publications. Delhi, India.

Kirkwood, R., Longley, A. 1995. *Clean Technology and the Environment.* 1st Ed. Springer. Dordrecht, The Netherlands.

Kitani, O. 1999. *CIGR Handbook of Agricultural Engineering. Energy Biomass and Engineering.* ASAE Publication. Saint Joseph, Missouri, US.

Klocke, F. 2009. *Manufacturing Processes. Grinding, Honing, Lapping.* 1st Ed. Springer-Verlag. Berlin, Germany.

Knight, J.B., Kotschevar, L.H. 2000. *Quantity: Food Production, Planning, and Management.* 3rd Ed. Wiley. Hoboken, New Jersey, US.

Knotts, T.L., Jones, S.C. Udell, G.G. 2009. Innovation Evaluation and Product Marketability. 19(2): 84-90.

Knutson, K., Schexnayder, C., Fiori, C., Mayo, R. 2008. *Construction Management Fundamentals.* 2nd Ed. McGraw-Hill. New York, US.

Koffka, K. 1962. *Principles of Gestalt Psychology.* 1st Ed. Lund Humphries. London, UK.

Kogon, K., Blakemore, S., Wood, J. 2015. *Project Management for the Unofficial Project Manager.* 1st Ed. BenBella. Texas, U.S.

Komleh, P.S., Omid, M., Keyhani, A., 2011. Study on energy use pattern and efficiency of corn silage in Iran by using data envelopment analysis (DEA) technique. *International Journal of Environmental Science.* 1(6): 1094-1106.

Kotler, P. 2002. *Marketing Management.* 11th Ed. Prentice Hall. New Jersey, US.

Kotler, P., Armstrong, G., Brown, L., Adam, S. 2006. *Marketing,* 7th Ed. Pearson Education Australia/Prentice Hall.

Kotter, J.P. 2012. *Leading Change.* With a New Preface by the Author. 1st Ed. Harvard Business Review Press. Massachusetts, US.

Koziol, H., Green, M.D., Lunney, M., Oliphant, K., Yang, L. 2018. *Production Liability: Fundamental Questions in a Comparative Perspective.* de Gruyter. Berlin, Germany.

Kroemer, A.D., Kroemer, K. 2016. *Office Ergonomics: Ease and Efficiency at Work.* 2nd Ed. CRC Press. Boca Raton, Florida, US.

Kroemer, K., Grandjean, E. 1997. *Fitting the Task to the Human.* 5th Ed. Taylor & Francis. Oxfordshire, UK.

Krugman, P., Wells, R. 2012. *Economics.* 3rd ed., Worth Publishers.

Krugman Paul, R., Obstfeld, M., Melitz, M. 2014. *International Economics: Theory and Policy.* 10th Ed. Pearson. New York, US.

Kuehn, C. 2015. *Multiple Time Scale Dynamics.* 1st Ed. Springer Nature. Brugg, Switzerland.

Kuhlman, T., Farrington, J. 2010. What is Sustainability? *Sustainability*, 2(11), 3436-3448.

Kumar, A., Prakash, O., Singh Chauhan, P., Gautam, S. 2020. *Energy Management Conservation and Audits. 1st* Ed. CRC Press. Florida, US.

Kumar Agariya, A., Johari, A., Sharma, H.K. Chandraul, U.N.S., Singh, D. 2012. The role of packaging in brand. *Communication International Journal of Scientific & Engineering Research.* 3(2): 1-13.

Kuper, A., Kuper, J. 2009. *The Social Science Encyclopedia.* 3rd Ed. Routledge. Oxfordshire, UK.

Kuper, A. Kuper, J. 2004. *The Social Science Encyclopedia.* 3rd Ed. Routledge. Oxford, UK.

Lance, J. 2013. *The Beginner's Guide to Engineering: Computer Engineering. CreateSpace Independent Publishing Platform.* California, US.

Lawrence, D.P. 2003. *Environmental Impact Assessment: Practical Solutions to Recurrent Problems.* 1st Ed. John Wiley & Sons. Hoboken, New Jersey, US.

Lawrence, F. 2013. *Strategy.* Oxford University Press. Oxford, UK.

Layton, M.C., Ostermiller, S.J., Kynaston, D.J. 2020. *Agile Project Management.* 3rd Ed. For Dummies. New Jersey, US.

Leal Filho, W., Surroop, D. 2018. *The Nexus: Energy, Environment and Climate Change.* Springer, New York City. New York, US.

Lefteri, C. 2012. *Making It: Manufacturing Techniques for Product Design.* 2nd Ed. Laurence King Publishing. London, UK.

Levitt, J. 2009. *Handbook of Maintenance Management* (Volume 1). 2nd Ed. Industrial Press. Connecticut, US.

Levitt, J.D. 2013. *Facilities Management: Managing Maintenance for Buildings and Facilities.* 1st Ed. Momentum Press. New York, US.

Lewis, L., Tietenberg, T.H. 2020. *Environmental Economics and Policy.* 7th Ed. Routledge. Oxfordshire, UK.

Lewis, R. 2015. *Muslim Fashion, Contemporary Style Cultures.* Duke University Press. North Carolina, US.

Li, E.-P. 2012. *Electrical Modeling and Design for 3D System Integration: 3D Integrated Circuits and Packaging, Signal Integrity, Power Integrity and EMC.* 1st Ed. Wiley-IEEE Press. New Jersey, US.

Liebowitz, J. 1998. *Information Technology Management: A Knowledge Repository.* 1st Ed. CRC Press. Florida, US.

Ling, C.H., Mansori, S. 2018. The Effects of Product Quality on Customer Satisfaction and Loyalty: Evidence from Malaysian Engineering Industry. *International Journal of Industrial Marketing.* 3(1): 20-35.

Little, W. 2016. *Introduction to Sociology,* 2nd Canadian Edition. BCcampus. Victoria, British Columbia, Canada.

Liu, G.R., Quek, S.S. 2003. *The Finite Element Method: A Practical Course.* 1st publication. Elsevier. Oxford, England.

Locke, E.A., Latham, G.P. 1990. *A theory of goal setting & task performance.* Prentice Hall, New Jersey, US.

Lohstroh, M. Derler, P., Sirjani, M. 2018. Principles of Modeling. *Essays Dedicated to Edward A. Lee on the Occasion of His 60th Birthday.* Springer, Berlin, Germany.

Lordan, E.J. 2003. *Essentials of Public Relations Management.* 1st Ed. Rowman & Littlefield Publishers. Maryland, US.

Lumsdaine, E., Binks, M. 2006. *Entrepreneurship from Creativity to Innovation: Effective Thinking Skills for a Changing World.* Trafford Publishing. Indiana, US.

Luthra, S., Garg, D., Agarwal, A., Mangla, S.K. 2020. *Total Quality Management* (TQM): Principles, Methods, and Applications. 1st Ed. CRC Press. Florida, US.

Lysons, K. 2016. *Procurement & Supply Chain Management.* 9th Ed. Trans-Atlantic Publications. Pennsylvania, US.

Machado, C., Davim, J.P. 2021. *Knowledge Management and Learning Organizations.* 1st Ed. Springer International Publishing. Cham, Switzerland.

Mackintosh, N. 2011. *IQ and Human Intelligence.* 2nd Ed. Oxford University Press. Oxford, UK.

MacLean-Blevins, M. 2017. *Designing Successful Products with Plastics.* 1st Ed. William Andrew. New York, US.

Magee, Liam; Scerri, Andy; James, Paul; Thom, James A.; Padghan, Lin; Hickmott, Sarah; Deng, Hepu; Cahill, Felicity. 2012. Reframing social sustainability reporting: towards an engaged approach. *Environment, Development and Sustainability.* 15(1): 225-243.

Mahesh, M., Kolla, S., Tirumala, K.M. 2018. Design Process to Reduce Production Cycle Time in Product Development. *IAES International Journal of Artificial Intelligence;* Yogyakarta. 7(3): 125-129.

Man, C.M.D., Jones, A.A. 1994. *Shelf Life Evaluation of Foods.* Springer, Berlin, Germany.

Mandour, M. 2019. *Culture and its organs.* Arab Press Agency. El Haram, Giza, Egypt Giza, Egypt. (In Arabic).

Manetti, G., Copenhaver, B.P. 2019. *On Human Worth and Excellence.* Harvard University Press. Massachusetts, US.

Manfredo, M.J., Vaske, J.J., Rechkemmer, A., Duke, E.A. 2014. *Understanding Society and Natural Resources.* Springer. Heidelberg, Germany.

Marco, J., Truong Dinh, Q., Longo, S. 2020. *Energy Storage and Management for Electric Vehicles.* 1st Ed. Mdpi Publication. Basel, Switzerland.

Markham, A.C. 2021. *A Brief History of Pollution.* 1st Ed. Routledge. Oxfordshire, UK.

Marolla, C. 2018. *Information and Communication Technology for Sustainable Development.* 1st Ed. CRC Press. Florida, Us.

Martin, P., Dantan, J.Y., Siadat, A. 2007. Cost Estimation and Conceptual Process Planning. In: Cunha P.F., Maropoulos P.G. *Digital Enterprise Technology.* 1st Ed. Springer, Boston, MA, US.

Martinez, D., Ebenhack, B., Wagner, T. 2019. *Energy Efficiency: Concepts and Calculations.* 1st Ed. Elsevier Science. Amsterdam, The Netherlands.

Mas-Colell, A., Whinston, M.D., Green, J.R. 1995. *Microeconomic Theory.* 1st Ed. Oxford University Press. Oxford, UK.

Massingham, P. 2019. *Knowledge Management: Theory in Practice.* 1st Ed. SAGE Publishing. London, UK.

May, A., Ross, T. 2018. The design of civic technology: factors that influence public participation and impact. *Ergonomics,* 61(2): 214-225.

McAfee, P., Lewis, T.R. 2009. *Introduction to Economic Analysis.* 1st Ed. Saylor Foundation. Washington, US.

Mcdavid, J.C., Huse, I., Ingleson, L.R.L. 2020. *Program Evaluation and Performance Measurement: An Introduction to Practice.* Volume 2. SAGE publication. Thousand Oaks, California, US.

McDonald, P.H. 2001. *Fundamentals of Infrastructure Engineering. Civil Engineering Systems.* 2nd Ed. CRC Press. New Jersey, US.

McKnight, A.D., Marstrand, P.K., Sinclair, T.C. 2021. *Environmental Pollution Control: Technical, Economic and Legal Aspects.* 1st Ed. Oxfordshire, UK.

Mehmood, Z., Haneef, I., Udrea, F. 2018. Material selection for Micro-Electro-Mechanical-Systems (MEMS) using Ashby's approach. *Materials & Design.* 157(5): 412-430.

Mehta, M.L., Verma, S.R., Misra, S.K. 2005. *Testing and Evaluation of Agricultural Machinery.* 1st Ed. Daya Publishing House. New Delhi, India.

Mehta, M.L., Verma, S.R., Mihra, S.K., Sharma, V.K. 2016. *Testing and Evaluation of Agricultural Machinery (Hardback).* Astral International Pvt. Dehli, India.

Merritt, F.S., Ricketts, J.T. 2000. Building Design and Construction Handbook. 6th Ed. McGraw-Hill Professional. New York, US.

Mersid Poturak, I.B. 2014. Influence of product packaging on purchase decisions. *Journal of Business Strategies.* 6(2): 1-10.

Meyer, M.D. 2016. *Transportation Planning Handbook.* 4th Ed. John Wiley & Sons. Hoboken, New Jersey, US.

References

Michaelides, E.E. 2018. *Energy, the Environment, and Sustainability.* 1st Ed. CRC Press. New Jersey, US.

Mishkin, F.S. 2007. *The Economics of Money, Banking, and Financial Markets.* 1st Ed. Boston: Addison Wesley. p. 8.

Michalowicz, M. 2017. *Profit First: Transform Your Business from a Cash-Eating Monster to a Money-Making Machine.* 1st Ed. Portfolio. New York, US.

Minkoff-Zern, L.-A. 2019. *The New American Farmer: Immigration, Race, and the Struggle for Sustainability* (Food, Health, and the Environment). 1st Ed. The MIT Press. Massachusetts, US.

Mildred, M. 2001. *Product Liability: Law and Insurance.* 1st Ed. Routledge, Oxforshire, UK.

Ming, B.S.Y. 2000. *Installation and commissioning of mechanical and electrical equipment.* 1st Ed. Beijing University of Aeronautics and Astronautics Press. Wudaokou, China.

Mirzaee Ghaleh, E., Rafiee, S., Keyhani, A.R., Emam Djom-eh, Z., Kheiralipour, K. 2008. Mass modeling of two varieties of apricot (prunus armenaica L.) with some physical Characteristics. *Journal of Plant Omics.* 1: 37-43.

Mobley, K.R. 2014. Maintenance Engineering Handbook. 8th Ed. McGraw-Hill Education. New York City, US. Mobley, R.K. 2002. *An Introduction to Predictive Maintenance.* 2nd Ed. Elsevier. Amsterdam, The Netherlands.

Mohammed, R., Murova, O. 2019. The Effect of Price Reduction on Consumer's Buying Behavior in the U.S. Differentiated Yogurt Market. *Applied Economics and Finance.* 6(2): 32-42.

Mohammadi, A., Tabatabaeefar, A., Shahin, S., Rafiee, S., Keyhani A. 2008. Energy use and economical analysis of potato production in Iran a case study: Ardabil province. *Energy Conservation and Management.* 49: 3566-3570.

Mohammadi, V., Kheiralipour, K., Ghasemi-Varnamkhasti, M. 2015. Detecting maturity of persimmon fruit based on image processing technique. *Scientia Horticulturae.* 184: 123-128

Mollah, H. Baseman, H., Long, M. 2013. *Risk Management Applications in Pharmaceutical and Biopharmaceutical Manufacturing.* 1st Ed. Wiley. New Jersey, US.

Monden, Y. 2011. *Toyota Production System: An Integrated Approach to Just-In-Time.* 4th Ed. Productivity Press. US. New York, US.

Monthei, D.L. 1999. *Package Electrical Modeling, Thermal Modeling, and Processing for GaAs Wireless Applications.* Springer. New York. US.

Moradian, S. 2018. *Manufacturer's civil liability in providing information about the goods.* 1st Ed. Ebtekar Ghalam. Tehran, Iran. (In Persian).

Morden, T. 2004. *Principles of Management.* 2nd Ed. Routledge.

Morris, B.A. 2016. *The Science and Technology of Flexible Packaging.* 1st Ed. Elsevier. Amsterdam, The Netherlands.

Moss, D., DeSanto, B. 2012. *Public Relations: A Managerial Perspectiv.* SAGE Publications. New York, US.

Murtagh, B. 2020. *Social Economics and the Solidarity City.* 1st Ed. Routledge. Oxfordshire, UK.

Muthu, S.S., Li, Y. 2013. *Manufacturing Processes of Grocery Shopping Bags. Assessment of Environmental Impact by Grocery Shopping Bags: An Eco-Functional Approach. Environmental Issues in Logistics and Manufacturing.* Springer Science & Business Media, Singapore.

Mullins, J., Walker, O. 2012. *Marketing Management: A Strategic Decision-Making Approach.* 8th Ed. McGraw-Hill. New York, US.

Murali Krishna, I.V., Manickam, V. 2017. *Environmental Management. Science and Engineering for Industry.* 1st Ed. Elsevier, The Netherlands.

Murthy, D.N.P., Blischke, W.R. 2006. *Warranty Management and Product Manufacture.* Springer-Verlag. London, UK.

Murthy, D.N.P., Jack N. 2003. Warranty and Maintenance. In: Pham H. (eds) *Handbook of Reliability Engineering.* Springer. London, UK.

Murthy, K.R.S., Kadur, A., Rao, P. 1994. A holistic approach product marketability measurements-the PMM approach. *IEEE.* Dayton North, Ohio, US.

Myers, J.G., 1975. *Energy Consumption in Manufacturing.* HarperCollins Distribution. Bishopbriggs, UK.

Myers, N., Spoolman, S. 2014. *Environmental Issues and Solutions: A Modular Approach.* 1st Ed. Cengage Learning. Boston, Massachusetts, US.

Nadakavukaren, A., Caravanos, J. 2020. *Our Global Environment: A Health Perspective.* 8th Ed. Waveland Press. Long Grove, Illinois, US.

Nagle, T.T., Müller, G. 2017. *The Strategy and Tactics of Pricing: A guide to growing more profitably.* 6th Ed. Routledge. Oxfordshire, UK.

Nanda, V. 2016. *Quality Management System Handbook for Product Development Companies.* CRC Press. New Jersey, US.

Nanay, B. 2019. *Aesthetics: A Very Short Introduction.* 1st Ed. OUP Oxford. Oxford, UK.

Nasseri, A. 2019. Energy use and economic analysis for wheat production by conservation tillage along with sprinkler irrigation. *Science of the Total Environment.* 648: 450-459.

Nee, A.Y.C. 2015. Handbook of Manufacturing Engineering and Technology. Springer, Berlin, Germany.

Nelson, V.C., Starcher, K.L. 2015. *Introduction to Renewable Energy.* 2nd Ed. CRC Press. New Jersey, US.

Nkomo, G.V., Sedibe, M.M., Mofokeng, M.A. 2021. Production constraints and improvement strategies of cowpea (*Vigna unguiculata* L. Walp.) genotypes for drought tolerance. *International Journal of Agronomy.* 2021(2): 1-9.

Noble, J. 2000. *Textbook of Primary Care Medicine.* 3rd edition. Mosby. Missouri, US.

Nongpluh, Y.S. 2013. *Know all about: reduce, reuse, and Recycle.* Noronha, Guy C., Energy and Resources Institute. New Delhi, India.

Noonan, C. 1998. *Sales Management.* 1st Ed. Routledge. Oxfordshire, UK.

Normann, R. 2001. *Service Management: Strategy and Leadership in Service Business.* 3rd Ed. Wiley. New Jersey, US.

Nuttall, J. 2002. *An Introduction to Philosophy.* 1st Ed. Polity. Cambridge, UK.

Ochsner, A., Altenbach, H. 2020. *Engineering Design Applications II.* 1st Ed. Springer Nature. Aktiengesellschaft, Switzerland.

Oliveira, J.A., Lopes Silva, D.A., Puglieri, F.N., Saavedra, Y.M.B. 2021. *Life Cycle Engineering and Management of Products: Theory and Practice.* Springer. Berlin, Germany.

Olsen, D. 2015. *The Lean Product Playbook: How to Innovate with Minimum Viable Products and Rapid Customer Feedback.* 1st Ed. Wiley. New York, US. Oumano, E. 2011. Cinema Today. Rutgers University Press. 1st Ed. New Jersey, US.

Olsen, E.K. 2017. *Labor and Labor Power. In: Routledge Handbook of Marxian Economics.* 1st Ed. Routledge. Oxfordshire, UK.

Owen, D. 2009. *Green Metropolis: What the City Can Teach the Country About True Sustainability.* Riverhead Hardcover. New York, US.

Ozolins, P. 2014. *Sustainability & Scarcity: A Handbook for Green Design and Construction in Developing Countries.* 1st Ed. Routledge. Oxfordshire, UK.

Pahl, G., Beitz, W., Feldhusen, J., Grote, K.-H. 2007. *Engineering Design, A Systematic Approach.* 1st Ed. Springer, London, UK.

Pahlavan, C. 2009. *Culture and civilization.* Nashr e Ney. Tehran, Iran. (In Persian).

Parker, D.H. 2012. *The Principles of Aesthetics.* CreateSpace Independent Publishing Platform. Scotts Valley, California, US.

Panneerselvam, R. 2013. *Engineering Economics.* 2nd Ed. Prentice Hall India Learning Private Limited. Delhi, India.

Parry, M.L. 2021. *Climate Change and World Agriculture.* 1st Ed. Oxfordshire, UK.

Pasher, E. 2011. *The Complete Guide to Knowledge Management.* 1st Ed. Wiley. New Jersey, US.

Patterson, D.A., Hennessy J.L. 2020. *Computer Organization and Design* MIPS Edition: The Hardware/Software Interface 6th Ed. Morgan Kaufmann. California, US.

Paulin, M.A. 2014. *The Idea Is...: A book for turning ideas into companies.* 1st Ed. CreateSpace Independent Publishing Platform. South Carolina, US.

Pavon, J. 2010. Human Attributes in the Modelling of Work Teams. In Proceedings of Balanced Automation Systems for Future Manufacturing Networks. *9th IFIP WG 5.5 International Conference,* July 21-23, Valencia, Spain.

Payandeh, Z., Kheiralipour, K., Karimi, M. 2016. Evaluation of energy efficiency of broiler production farms using data envelopment analysis technique, case study: Isfahan Province. *Iranian Journal of Biosystems Engineering,* 47(3), 577-585.

Payandeh, Z., Kheiralipour, K., Karimi, M., & Khoshnevisan, B. 2017. Joint data envelopment analysis and life cycle assessment for environmental impact reduction in broiler production systems. *Energy,* 127, 768-774.

Pena, L. 1991. *The Boundary between Scientific and Non-Scientific Knowledge.* 4th Annual Meeting of SOFIA. Salamanca, Spain.

Pettinger, R. 2006. *Introduction to Management.* 4th Ed. Red Globe Press. Gordonsville, Virginia, US.

Perez-Gosende, P., Mula, J., Diaz-Madronero, M. 2021. Facility layout planning. An extended literature review. *International Journal of Production Research.* 59(12): 3777-3816.

Perez-Uribe, R., Salcedo-Perez, C., Ocampo-Guzman, D. 2018. *Handbook of Research on Entrepreneurship and Organizational Sustainability in SMEs* (Advances in Logistics, Operations, and Management Science). 1st Ed. IGI Global. Pennsylvania, US.

Pernick, R., Wilder, C. 2007. *The Clean Tech Revolution: The Next Big Growth and Investment Opportunity.* 1st Ed. Harper Business. New York City, New York, US.

Perroux, F. 2010. *A New Concept of Development, Basic Tenets.* 1st Ed. Routledge, London, UK.

Peters, T.J. 2005. *Design: Tom Peters Essentials.* 1st Ed. Dorling Kindersley. London, UK.

Pezzullo, P.C., Cox, R. 2021. *Environmental Communication and the Public Sphere.* 6th Ed. SAGE Publishing. Thousand Oaks, California, US.

Pfaffenberger, B. 1992. *Social Anthropology of Technology Annual Review of Anthropology.* 21: 491-516.

Pfeiffer, F., Bremer, H. 2017. *The Art of Modeling Mechanical Systems.* Springer International Publishing, Berlin, Germany.

Pham, H. 2003. *Handbook of Reliability Engineering.* 1st Ed. Springer-Verlag. London, UK.

Pidd, M. 1996. Five Simple Principles of Modelling. *IEEE.* California, US.

Pine, T., Lee, M.M.K., Jones, T.B. 1999. Weight reduction in automotive structures-an experimental study on torsional stiffness of box sections. *Journal of Automobile Engineering.* 213(1): 59-71.

Popovici, K., Mosterman, P.J. 2017. *Real-Time Simulation Technologies: Principles, Methodologies, and Applications.* 1st Ed. CRC Press. Florida, US.

Potschin, M., Haines-Young, R. 2008. Sustainability impact assessments: Limits, thresholds and the sustainability choice space. In *Sustainability Impact Assessment of Land Use Changes;* Helming, K., Pérez-Soba, M., Tabbush, P., Eds.; Springer: Berlin, Germany, 2008; pp. 425-450.

Prada, C.D., Pantelides, C. Pitarch, J.L. 2019. *Process Modelling and Simulation.* Processes, MDPI. Basel, Switzerland.

Prasad, J., Nair, C.G.K. 2011. *Non-Destructive Test and Evaluation of Materials.* McGraw Hill Education. New York City, US.

Pristavka, M., Kotorova, M., Savov, R. 2016. Quality Control in Production Processes. *Acta Technologica Agriculturae.* 19(3):77-83.

Priyono, P.I. 2017. Effect of Quality Products, Services and Brand on Customer Satisfaction at McDonald's. *Journal of Global Economics.* 5(2): 1000247.

Ptak, C.A., Smith, C. 2011. *Orlicky's Material Requirements Planning.* 3rd Ed. McGraw-Hill. New York, US.

Pyzdek, T., Keller, P. 2013. *The Handbook for Quality Management, Second Edition: A Complete Guide to Operational Excellence.* 2nd Ed. McGraw-Hill. New York, US.

Radforld, J.D. 1980. *Production Engineering Technology* 3rd Ed. Palgrave. London, United Kingdom.

Radojicic, M., Nesic, Z., Vasovic, J.V. 2013. Characteristics of the impact of production volume on cost dynamics and unit cost of products. *Metalurgia International.* 18: 236-241.

Raj, B., Jayakumar, T., Thavasimuthu, M., Karamanis, N., Artincz-Botas, R.F. 2002. *Practical nondestructive testing.* 1st Ed. Woodhead Publishing. Cambridge, UK.

Rajput, R.K. 2017. *A Textbook of Manufacturing Technology: Manufacturing Processes.* 2nd Ed. Laxmi Publications. New Delhi, India.

Ramedani, Z., Alimohammadian, L., Kheialipour, K., Delpisheh, P., & Abbasi, Z. (2019). Comparing energy state and environmental impacts in ostrich and chicken production systems. *Environmental Science and Pollution Research.* 26(27): 28284-28293.

Rasmussen, S. 2013. *Production Economics. The Basic Theory of Production Optimization.* Springer. Berlin, Germany.

Rastogi, P.N. 2009. *Management of Technology and Innovation: Competing Through Technological Excellence.* 2nd Ed. SAGE Publications. London, UK.

Ray, P.K., Maiti, J. 2018. *Ergonomic Design of Products and Worksystems - 21st Century Perspectives of Asia.* Springer. Gateway East, Singapore.

Reed, K.B., Tech, V. 2020. *Strategic Management.* Virginia Tech Publishing.

Reisch, L.A., Thogersen, J. 2015. *Handbook of Research on Sustainable Consumption.* 1st Ed. Edward Elgar Publishing. Cheltenham, Gloucestershire, UK.

Render, B., Munson, C., Sachan, A., Heizer, J. 2018. *Operations Management Sustainability and Supply Chain Management.* 12th Ed. Pearson India. Delhi, India.

Reniers, G.L.L., Sorensen, K., Vrancken, K. 2013. *Management Principles of Sustainable Industrial Chemistry: Theories, Concepts and Indusstrial Examples for Achieving Sustainable Chemical Products and Processes from a Non-Technological Viewpoint.* 1st Ed. Wiley-VCH. New Jersey, US.

Reingruber, M.C., Gregory, W.W. 1994. *The Data Modeling Handbook: A Best-Practice Approach to Building Quality Data Models.* 1st Ed. Wiley. New Jersey, US.

Renna, P., Ambrico, M. 2021. *Design and Optimization of Production Lines.* Applied Sciences. MDPI. Basel, Switzerland.

Richards, H. 2020. *Grow Food For Free: The sustainable, zero-cost, low-effort way to a bountiful harvest.* DK. London, UK.

Rima, I.H. 2009. *Development of Economic Analysis* 1st Ed. Routledge. Oxfordshire, UK.

Robert T. Kiyosaki, R.T. 2011. *Rich Dad's CASHFLOW Quadrant: Rich Dad's Guide to Financial Freedom.* 2nd Ed. Plata Publishing. Beijing, China.

Robinson, M. 2018. *Climate Justice. Hope, Resilience, and the Fight for a Sustainable Future.* 1st Ed. Bloomsbury Publishing. London, UK.

Rogoff, M.J. 2013. *Solid Waste Recycling and Processing. Planning of Solid Waste Recycling Facilities and Programs.* 2nd Ed. William Andrew. New York, US.

Roper, K., Payant, R. 2014. *The Facility Management Handbook.* 4th Ed. AMACOM. New York, US.

Ross, P.H., Ellipse, M.W., Freeman, H.E. 2004. *Evaluation: A systematic approach.* 7th Ed. Thousand Oaks: Sage.

Rosen, L.D., Cheever, N.A., Carrier, L.M. 2015. *The Wiley Handbook of Psychology, Technology, and Society.* John Wiley & Sons. Hoboken, New Jersey, US.

Rosenberg, M. 2015. *Strategy and Sustainability: A Hardnosed and Clear-Eyed Approach to Environmental Sustainability for Business* (IESE Business Collection). 1st Ed. Springer. Berlin, Germany.

Roser, M., Ritchie, H. 2020. *Environmental impacts of food production.* Our world in data.

Rouhani, A., Keshavarz, A. 2014. *Technology Management.* Aseman Negar publication. Esfahan, Iran.

Rouillard, L. 2003. *Goals and Goal Setting: Achieving Measured Objectives.* 1st Ed. Crisp Publications, Hamilton, US.

References

Rumane, A.R. 2019. *Quality Management in Construction Projects.* 2nd Ed. CRC Press. Florida, US.

Sackrey, C., Schneider, G., Knoedler, J. 2013. *Introduction to Political Economy.* 7th Ed. Dollars & Sense. Massachusetts, US.

Sahota, A. 2014. *Sustainability: How the Cosmetics Industry is Greening Up.* 1st Ed. Wiley. New Jersey, US.

Salvatore, A. 2016. *The sociology of Islam: knowledge, power and civility.* 1st Ed. Wiley Blackwell. Hoboken, New Jersey, US.

Sasvari, P. 2012. The Effects of Technology and Innovation on Society. *Bahria University Journal of Information & Communication Technology.* 5(1): 1-10.

Sammut-Bonnici, T. 2015. Brand and Branding. In: Wiley Encyclopedia of Management. Vol 12. Strategic Management. John Wiley & Sons.

Samuelson, P. 1983. *Foundations of Economic Analysis.* 1st Ed. Harvard University Press. Massachusetts, US.

Sandfort, J., Moulton, S. 2015. *Effective Implementation in Practice: Integrating Public Policy and Management.* Jossey-Bass. New Jersey, US.

Sayigh, A.A.M. 1990. *Energy and the Environment.* 1st Ed. Pergamon. Oxford, UK.

Schellens, M.K., Gisladottir, J. 2018. Critical Natural Resources: Challenging the Current Discourse and Proposal for a Holistic Definition. *Resources,* 7(4): 79.

Schmitt, B., Simonson, A. 2009. *Marketing Aesthetics.* 1st Ed. The Free Press. New York, US.

Schubert, P. 2000. The Participatory Electronic Product Catalogue: Supporting Customer Collaboration in E-Commerce Applications. *Electronic Markets* 10(4): 229-236.

Schumpeter, J.A. 1955. *History of Economic Analysis.* 1st Ed. Routledge. Oxfordshire, UK.

Sclater, N. 2003. *Handbook of Electrical Design Details.* 2nd Ed. McGraw-Hill Education. New York, US.

Scorer, R.S. 2021. *Pollution in the Air: Problems, Policies and Priorities.* 1st Ed. Oxfordshire, UK.

Selikoff, S. 2020. *The Complete Book of Product Design, Development, Manufacturing, and Sales.* 2nd Ed. Product Development Academy, *Washington State, USA*.

Sellars, W. 1963. *Empiricism and the Philosophy of Mind.* Routledge and Kegan Paul Ltd. Oxfordshire, UK.

Seufert, E.B. 2014. Freemium Monetization. Chapter 6. *Freemium Economics. Leveraging Analytics and User Segmentation to Drive Revenue.* Elsevier, Amsterdam, The Netherlands.

Seubert, T., Vokey, G. 2020. *Manufacturing Execution Systems: An Operations Management Approach.* 1st Ed. ISA - International Society of Automation. North Carolina, US.

Sharma, H.D., Reddy, K.R. 2004. *Geoenvironmental Engineering: Site Remediation, Waste Containment, and Emerging Waste Management Techonolgies.* 1st Ed. Wiley. New Jersey, US.

Sharma, P. C. 2006. *A Textbook of Production Technology: Manufacturing Processes.* 1st Ed. S. Chand. Delhi, India.

Sharma, D.N., Mukesh, S. *Farm Machinery Design, Principles and Problems.* 3rd Ed. Delhi, India.

References

Sharma, K.L.S. 2016. *Overview of Industrial Process Automation.* 2nd Ed. Elsevier. Amsterdam, The Netherland.

Shearlock, C. 2019. *Storing Food without Refrigeration.* Blue River Press. Indianapolis, Indiana, US.

Shigley, J.E., Mischke, C.R., Brown Jr. T.H. 2004. *Standard Handbook of Machine Design.* 3rd Ed. McGraw-Hill Educatio, New York, US.

Shinsuke, M., Masahiro, O., Tomohiro, T., Ichiro, D., Seiji, H. 2010. Lifespan of Commodities, Part I. *Journal of Industrial Ecology.* 14 (4): 598-612.

Shitole, G.Y., Sable, R. 2012. *Environmental Degradation Issues and Challenges.* 1st Ed. Serials Publication. New Delhi, Delhi, India.

Shiva, V. 2005. *Earth Democracy: Justice, Sustainability, and Peace.* South End Presslocation. York City, US.

Sickles, R., Zelenyuk, V. 2019. *Measurement of Productivity and Efficiency: Theory and Practice.* Cambridge University Press. Cambridge, UK.

Silver, N. 2017. *Finance, Society and Sustainability: How to Make the Financial System Work for the Economy, People and Planet.* 1st Ed. Springer. Berlin, Germany.

Siebert, H. 2008. *Economics of the Environment. Theory and Policy.* Springer-Verlag. Berlin, Germany.

Silva Chaves, E.D.P. 2017. Identity, Positioning, Brand Image and Brand Equity Comparison. *Independent Journal of Management & Production* 8(4):1246.

Simon, H., Fassnacht, M. 2019. 1st Ed. *Price Management: Strategy, Analysis, Decision, Implementation.* Springer. Berlin, Germany.

Singh, J. 2011. *Public Relations Management: Practical Guidelines for Effective PR Management.* 1st Ed. Global India Publications. Delhi. India.

Sioshansi, F. 2011. *Energy, Sustainability and the Environment.* 1st Ed. Butterworth-Heinemann. Oxford, UK.

Skakoon, J.G. 2008. *The Elements of Mechanical Design.* 1st Ed. ASME Press. New York, US.

Skipka, K.J., Theodore, L. 2014. *Energy Resources: Availability, Management, and Environmental Impacts.* 1st Ed. CRC Press. New Jersey, US.

Smith, C.B., Parmenter, K. 2015. *Energy Management Principles. Applications, Benefits, Savings.* 2nd Ed. Elsevier. The Netherland.

Smith, T. 2011. *Pricing Strategy: Setting Price Levels, Managing Price Discounts and Establishing Price Structures.* 1st Ed. Cengage Learning. Massachusetts, US.

Snow, R.E. Federico, P.-A., Montague W.E. 2021. A*ptitude, Learning, and Instruction:* Volume 1: Cognitive Process Analyses of Aptitude. 1st Ed. Routledge. Oxfordshire, UK.

Snow, R.E. Federico, P.-A., Montague W.E. 2021. A*ptitude, Learning, and Instruction:* Volume 2: Cognitive Process Analyses of Learning and Problem Solving. 1st Ed. Routledge. Oxfordshire, UK.

Snow, R.E., Farr, M.J. 2021. *Aptitude, Learning, and Instruction:* Volume 3: Conative and Affective Process Analyses. 1st Ed. Routledge. Oxfordshire, UK.

Sokolowski, J.A., Banks, C.M. 2008. *Principles of Modeling and Simulation: A Multidisciplinary Approach.* 1st Ed. Wiley. New Jersey, US.

Sonnemann, G., Manuele MargnI, M. 2015. *Life Cycle Management* (LCA Compendium-The Complete World of Life Cycle Assessment). Springer. Berlin, Germany.

Sowell, T. 2014. *Basic Economics.* 5th Ed. Basic Books. New York, US.

Spasojevic Brkic, V., Klarin, M., Stanisavljev, S., Brkic, A., Sajfert, Z. 2016. Reduction of production cycle time by optimising production and non-production components of time in the metalworking industry: A case study. *South African Journal of Industrial Engineering.* 27(1): 178-191.

Spellman, F.R. 2014. *Environmental Impacts of Renewable Energy.* 1st Ed. CRC Press. New Jersey, US.

Squires, S. 2021. *Anthropology, Technology, and Innovation.* Oxford University Press. Oxford, UK.

Srivastava, A.K. 2005. *Engineering Principles of Agricultural Machines.* 2nd Ed. merican Society of Agricultural Engineers. St. Joseph, US.

Stanton, D. 2020. *Supply Chain Management.* 1st Ed. For Dummies. New Jersey, US.

Stanton, N.A., Young, M.S., Harvey, C. 2014. *A Guide to Methodology in Ergonomics: Designing for Human Use.* 2nd Ed. CRC Press. Boca Raton, Florida, US.

Stenn, T.L. 2017. *Social Entrepreneurship as Sustainable Development: Introducing the Sustainability Lens.* 1st Ed. Palgrave Macmillan. London, UK.

Stiglitz, J.E. 1997. Reply-georgescu-roegen versus solow/stiglitz. *Ecol. Econ.* 22, 269–270.

Storey, J. 2021. *Cultural Theory and Popular Culture: An Introduction.* 9st Ed. Routledge. Oxfordshire, UK.

Stredwick, J. 2014. *An Introduction to Human Resource Management.* 3rd Ed. Routledge. Oxfordshire, UK.

Subramanian, M.N. 2019. *Plastics Waste Management.* 1st Ed. Scrivener Publishing LLC. Massachusetts, US.

Suchman, L. 2011. Anthropological Relocations and the Limits of Design. *Annual Review of Anthropology.* 40(1): 1-18.

Sujova, A., Marcinekova, K., Hittmar, S. 2017. Sustainable Optimization of Manufacturing Process Effectiveness in Furniture Production. *Sustainability.* 9: 923.

Sule, D.R. 2008. *Manufacturing Facilities: Location, Planning, and Design.* 3rd Ed. CRC Press. Boca Raton, Florida, US.

Sule, D.R. 2007. *Production Planning and Industrial Scheduling: Examples, Case Studies and Applications.* 2nd Ed. CRC Press. Florida, US.

Suzuki, H., Cervero, R., Iuchi, K. 2013. *Transforming Cities with Transit: Transit and Land-Use Integration for Sustainable Urban Development.* World Bank Publications. Virginia, US.

Taliaferro, C. 2011. *Aesthetics: A Beginner's Guide.* 1st Ed. Oneworld Publications. London, UK.

Tan, W. 2017. *Research Methods: A Practical Guide for Students and Researchers.* 1st Ed. WSPC. Paleo Faliro, Greece.

Tanchoco, J.M. 1994. *Material Flow Systems In Manufacturing.* Springer. New York, US.

Tang, H. 2017. *Automotive Vehicle Assembly Processes and Operations Management.* SAE. Warrendale, Pennsylvania, US.

Taylor, T., Greenlaw, S.A., Dodge, E. 2014. *Principles of Microeconomics.* 1st Ed. OpenStax. Texas, US.

References

Tchobanoglous, G., Kreith, F. 2002. *Handbook of Solid Waste Management.* 2nd Ed. McGraw-Hill. New York, US.

Tempelman, E. Shercliff, H., van Eyben, B.N. 2014. *Manufacturing and Design: Understanding the Principles of How Things Are Made.* 1st Ed. Heinemann. New Hampshire, US.

Tennent, J. 2014. *The Economist Guide to Financial Management: Principles and practice.* 2nd Ed. The Economist. Missouri, US.

Thiede, S. 2012. *Energy Efficiency in Manufacturing Systems.* 1st Ed. Springer-Verlag. Berlin, Germany.

Thomas, A.M. 2021. *Macroeconomics. An Introduction.* 1st Ed. Cambridge University Press. Cambridge, UK.

Thompson, R. 2007. *Manufacturing Processes for Design Professionals.* 1st Ed. Thames & Hudson, London, UK.

Thumann, A., Franz, H. 2009. *Efficient Electrical Systems Design Handbook.* River Publishers. Gistrup, Denmark.

Theodore, M.K., Theodore, L. 2009. *Introduction to Environmental Management.* 1st Ed. CRC Press. Florida, US.

Thilmany, D., Bond, C., Keeling Bond, J. 2008. Understanding Consumer Interest in Product and Process-Based Attributes for Fresh Produce. *Agribusiness.* 24(2): 231-252.

Thiry, M. 2010. *Program Management* (Fundamentals of Project Management). 1st Ed. Gower. Surrey, UK.

Timings, R. 2008. *Fabrication and Welding Engineering.* 1st Ed. Routledge. Oxfordshire, UK.

Tlusty, G. 1999. *Manufacturing Process and Equipment.* 1st Ed. Pearson. New Jersey, US.

Todaro, M.P. 2014. *Economic Development.* 12th Ed. Trans-Atlantic Publications. Philadelphia, US.

Tonn, B.E. 2007. Futures sustainability. *Futures.* 39, 1097–1116.

Trine, K.-M., Christini, D.J. 2020. *Modeling and Simulating Cardiac Electrical Activity.* IOP Publishing. Bristol, UK.

Tylor, E. 1871. *Primitive Culture.* 1st Ed. J.P. Putnam's Son. New York, US.

Tyson, S. 2015. *Essentials of Human Resource Management.* 6th Ed. Routledge. Oxfordshire, UK.

Ugural, A.C. 2020. *Mechanical Engineering Design.* 3rd Ed. CRC Press. Florida, US.

Ullman, D.G. 2009. *The Mechanical Design Process.* 4th Ed., Mc Graw Hill.

Ulrich, D., Smallwood, N. 2013. *Leadership Sustainability: Seven Disciplines to Achieve the Changes Great Leaders Know They Must Make.* McGraw-Hill Education. New York, US.

Ulrich, K.T. 2011. *Product Design and Development.* 5th Ed. McGraw-Hill Education. New York, US.

Vahab, A.A. 2010. *Introduction to Islamic Psychology.* Genuine Publications. Nizamuddin West, New Delhi, India.

van de Poel, I.R. 2020. Three philosophical perspectives on the relation between technology and society, and how they affect the current debate about artificial intelligence. *Human Affairs.* 30(4): 499-511.

References

Van de Putte, A., Kelimbetov, K., Holder, A. 2017. *The Perfect Storm: Navigating the Sustainable Energy Transition.* CreateSpace Independent Publishing Platform. California, US.

van Maarseveen, M., Martinez, J., Flacke, J. 2019. *GIS in Sustainable Urban Planning and Management: A Global Perspective.* 1st Ed. CRC Press. Florida, US.

Van Weele, A.J. 2018. *Purchasing and Supply Chain Management.* 7th Ed. Cengage Learning Business Press. Massachusetts, US.

Varshney, A. C. 2004. Data Book for Agricultural Machinery Design. 1st Ed. Central Institute of Agricultural Engineering, Madhya Pradesh, India. Vernon, D. 2009. *Human Potential: Exploring Techniques Used to Enhance Human Performance.* 1st Ed. Routledge. Oxfordshire, UK.

Veseth, M. 1984. *Introductory Macroeconomics.* 2nd Ed. Academic Press. Massachusetts, US.

Vitorio Andreoli, C., von Sperling, M., Fernandes, F. 2007. *Sludge Treatment and Disposal.* 1st Ed. IWA Publishing.

Wadhawan, V. 2017. *Understanding Natural Phenomena: Self-Organization and Emergence in Complex Systems.* 1st Ed. CreateSpace Independent Publishing Platform. Scotts Valley, California, US.

Waite, R. 2013. *Household Waste Recycling.* 1st Ed. Routledge. Oxfordshire, UK.

Walker, N. 2018. Sustainability: A Love Story (21st Century Essays). *Mad Creek Books.* Ohio, US.

Walker, S., Giard, J., Walker, H. 2017. *The Handbook of Design for Sustainability.* Bloomsbury Academic. London, UK.

Wang, Y. 2020. *The Handbook of Natural Resources,* Second Edition, Six Volume Set. 1st Ed. CRC Press. Florida. US.

Wasson, C.S. 2005. *System Analysis, Design, and Development: Concepts, Principles, and Practices.* 1st Ed. Wiley. New Jersey, US.

Weber, G. 2018. *Sustainability and Energy Management. Innovative and Responsible Business Practices for Sustainable Energy Strategies of Enterprises in Relation with CSR.* Springer. Wiesbaden, Germany.

Weinberg, M., Ganser, L.J. Studios, A. 2015. *Sales Management. Simplified: The Straight Truth About Getting Exceptional Results from Your Sales.* Audible Studios. New Jersey, US.

Welbourne, M. 2001. *Knowledge.* 1st Ed. Routledge. Oxfordshire, UK.

Werbach, A. 2009. *Strategy for Sustainability: A Business Manifesto.* Harvard Business Press. Massachusetts, US.

WCED. 1987. World Commission on Environment and Development (WCED). *Our Common Future;* Oxford University Press: New York, NY, USA, 1987.

Weske, M. 2012. *Business Process Management: Concepts, Languages, Architectures.* 3rd Ed. Springer-Verlag Berlin Heidelberg, Germany.

Wheeler, D.J. 2010. *Reducing Production Costs.* 1st Ed. SPC Press. Tennessee, US.

Whitbeck, R.H. 1934. *Industrial geography: Production, manufacture, commerce.* American Book Company. Knoxville, Tennessee, US.

Wiersum, K.F. 1995. 200 Years of Sustainability in Forestry: Lessons from History. *Environ. Manage.* 19, 321-329.

References

Wilderer, P.A. 2007. Sustainable water resource management: The science behind the scene. *Sustain. Sci.*, 2, 1-4.

Wilkinghoff, S. 2009. *Found Money*. 1st Ed. Wiley. New York, US.

Williams, A.C., Albertson, J.D. 2005. Contrasting short- and long-timescale effects of vegetation dynamics on water and carbon fluxes in water-limited ecosystems. *Water Resources Research*. 41(6): W06005.

Williams, J.C., Wright, B.D. 1991. *Storage and Commodity Markets*. Cambridge Press. Cambridge, UK.

Willis, H.E. 1926. A Definition of Law. *Virginia Law Review*. 12 (3): 203-214.

Wilson, J.R., Corlett, N. 2005. *Evaluation of Human Work*. 3rd Ed. CRC Press. Boca Raton, Florida, US.

Wilson, T.C. 2015. *Value and Capital Management: A Handbook for the Finance and Risk Functions of Financial Institutions*. 1st Ed. Wiley. New Jersey, US.

Wirtz, J. 2017. *Developing Service Products and Brands*. World Scientific. Singapore.

Woollard, F.G., Emiliani, B. 2009. *Principles of Mass and Flow Production*. 1st Ed. Clbm, LLC. New Jersey, US.

Worrell, E., Reuter, M. 2014. *Handbook of Recycling*. 1st Ed. Elsevier. Amsterdam, The Netherlands.

Wouters, M., Selto, F.H., Hilton, R.W., Maher, M.W. 2012. *Cost Management: Strategies for Business Decisions*, 1st Ed. McGraw-Hill. New York, US.

Xiao, J.J. 2019. *Consumer Economic Wellbeing*. 1st Ed. Springer. New York, US.

Yang, K., Xiaohua, J. 2014. *Security for Cloud Storage Systems*. Springer-Verlag New York, US.

Yeatman, A. 1996. The roles of scientific and non-scientific types of knowledge in the improvement of practice. *Australian Journal of Education*. 40(3): 248-30.

Yoo, M., Glardon, R. 2018. *Manufacturing Operations Management*. 1st Ed. World Scientific Publishing Europe. London, UK.

Youssef, H.A. El-Hofy, H.A. Ahmed, M.H. 2012. *Manufacturing Technology Materials, Processes, and Equipment*. 1st Ed. CRC Press. New Jersey, US.

Zandvakili, M. 2010. *Theories of Islamic Sociology*. Bakhshayesh Publishing. Tehran, Iran. (In Persian).

Zangeneh, M., Omid, M., Akram, A. 2010. A comparative study on energy use and cost analysis of potato production under different farming technologies in Hamadan province of Iran. *Energy*. 35(7): 2927-2933.

Zeigler, B.P., Muzy, A., Kofman, E. 2018. *Theory of Modeling and Simulation: Discrete Event & Iterative System Computational Foundations*. 3rd Ed. Academic Press. Massachusetts, US.

Zenios, S.A., Ziemba, W. 2007. Handbook of Asset and Liability Management Volume 2: *Applications and Case Studies*. 1st Ed. North-Holland publication. North-Holland, the Netherlands.

Zhu, L., Li, N., Childs, P.R.N. 2018. Light-weighting in aerospace component and system design. *Propulsion and Power Research*. 7(2): 103-119.

Zouelm, A., 2018. *Lean Islamic Culture: Fundamentals and Conceptual Patterns*. Publishing Organization of the Institute of Islamic Culture and Thought. Tehran, Iran. (In Persian).

About the Author

Kamran Kheiralipour
Mechanical Engineering of Biosystems
Ilam University
Pajhouhesh St., Ilam, Iran

Index

A

accountability, 30
accounting, 36
acquaintance, 34
adjustment, 12
aerospace, 114
aesthetics, 69, 75
agencies, 73
agricultural economics, 49
agriculture, 1, 2, 6, 42, 47, 71, 85, 99
algorithm, 21
anthropology, 63, 64, 67
aptitude, 25
aquifers, 61
artificial intelligence, 112
Asia, 108
assessment, 37, 44, 56, 57, 79, 95, 98, 99, 100, 106
assets, 36, 43
atmosphere, 55
attitudes, 64
audits, 38
automate, 32
Automobile, 107
automobiles, 4, 77
awareness, 26, 34

B

banking, 2, 36
base, 30, 36
batteries, 79
behaviors, 16, 68
Beijing, 104, 108

benefits, 8, 9, 33, 53, 61, 90
beverages, 79
biodiesel, 26, 60, 100
biogas, 26, 60
biomass, 60, 62
blends, 100
brain, 66
brain functioning, 66
brand image, 71, 81
business environment, 92
business management, 31
business processes, 11
buyer, 50
buyers, 82, 83

C

CAE, 90
carbon, 114
case study, 94, 100, 104, 106, 111
cash, 27, 36
cash flow, 27
chemical, 2, 5, 26
chemicals, 26, 79
chicken, 108
China, 104, 108
cities, 42
citizens, 66
citizenship, 66
civilization, 9, 43, 106
classification, 100
cleaning, 4, 61, 72
clients, 32

climate, 45, 55, 56, 65
climate change, 56
clothing, 11, 64
clusters, 4
CMC, 96
coal, 61
coding, 11
cognition, 66
color, 72, 75
commerce, 113
commodity, vii, viii, 1, 4, 5, 6, 7, 9, 11, 12, 19, 21, 22, 23, 32, 45, 47, 49, 50, 51, 53, 62, 68, 69, 73, 74, 76, 78, 79, 80, 81, 82, 83, 84, 114
communication, 14, 17, 46
communication skills, 17
community, 42
compensation, 33
competition, 34, 70, 83
competitive markets, 94
competitiveness, 31
competitors, 71
complement, 42
complexity, ix
composites, 95
computational fluid dynamics, 14
computer, 2, 5, 7, 12, 14
computer software, 12
computer-aided design, 14
computer-aided design (CAD), 14
computer-aided engineering, 14
computer-aided engineering (CAE), 14
conceptual model, 94
conceptualization, 16
configuration, 35
conflict, 30, 43
conflict resolution, 30
Congress, 99
conservation, 44, 105
construction, ix, 22, 32
consumers, vii, viii, 2, 6, 9, 13, 26, 31, 32, 35, 37, 38, 41, 61, 64, 71, 73, 74, 75, 78, 79, 81, 82, 83
consumption, 6, 35, 42, 45, 47, 49, 50, 58, 59, 60, 65

containers, 80
cooking, 64
cooling, 6, 24
cooperation, 26, 67
corrosion, 71
corruption, 81, 82, 83
cost, viii, 9, 14, 17, 29, 31, 39, 43, 49, 50, 51, 52, 53, 70, 75, 76, 78, 81, 107, 108, 114
cost-benefit analysis, viii, 43, 50
creativity, 14
crises, 37
criticism, 18
crops, 2, 71
CSR, 113
cultural heritage, 43
culture, 9, 64, 66, 67
customers, 4, 9, 18, 37, 50, 63, 64, 65, 72, 81, 83, 84

D

damages, 37, 82
danger, 73
data collection, 87
DEA, 101
debts, 27
decision-making process, 11, 13
defects, 37, 72, 73
Denmark, 112
depreciation, 81
derivatives, 61
design importance, ix, 11, 12
design process, ix, 11, 19, 50, 102, 112
design requirements, ix, 11, 19
design types, ix, 11, 13
designers, 12, 14
diabetes, 45
diesel fuel, 75
diet, 45
disability, 50
disappointment, 9
discomfort, 15
discriminant analysis, 93
disgust, 25
dissatisfaction, 15

Index

distribution, 4, 49, 82
diversity, 43
drawing, 12, 14, 17
drought, 105
drugs, 79

E

economic efficiency, 49, 50, 53
economic losses, 68
economic power, 8, 65
economics, 49, 65, 97
ecosystem, 43
education, 2, 16, 34, 65, 66
Egypt, 103
electric charge, 55
electricity, 15, 17, 27, 60
emission, 3, 5, 9, 58, 59
emotion, 66
employability, 30
employees, 30
employment, 9, 39
encouragement, 33
endangered, 45
energy, 1, 3, 5, 23, 26, 27, 29, 36, 38,
    42, 47, 51, 55, 56, 58, 59, 60, 61, 62,
    67, 68, 69, 70, 71, 72, 75, 78, 80, 81,
    82, 83, 87, 90, 94, 100, 101, 106,
    108, 114
energy consumption, 42, 58, 59, 60, 70, 72
energy efficiency, 59, 60, 106
energy input, 27
energy recovery, 62
enforcement, 67
engineering, 2, 8, 11, 13, 14, 31, 33, 49, 100
England, 102
entrepreneurship, 3, 46
environment, viii, ix, 2, 3, 4, 9, 38, 39,
    43, 44, 45, 50, 55, 56, 62, 76, 81, 82, 90
environmental aspects, vii, 35, 41, 44,
    45, 46, 51, 55, 56, 72, 76
environmental degradation, 55
environmental economics, 49
environmental effects, 38, 56
environmental impact, 45, 55, 56, 57,
    58, 59, 62, 69, 100, 106, 108
environmental management, 38
environmental protection, 42
environmental sustainability, 58, 80, 81
equilibrium, 39, 42, 47, 56
equipment, 5, 24, 29, 31, 36, 70, 74, 77, 79, 104
equity, 82
ethanol, 100
ethics, 14, 64, 66, 73, 83
Europe, 114
everyday life, 64
expertise, 30
explosives, 79
extraction, 2, 3, 6, 7

F

Fabrication, 98, 112
factories, vii, 6, 26, 36, 61, 67, 68
farms, 57, 96, 106
fat, 45
feelings, 30, 66
FEM, 17
fertility, 41
financial, 31, 34, 35, 36, 67, 71, 81
financial planning, 36
finite element method, 14, 17
fixed costs, 27, 51
flaws, 37
flexibility, 38, 71
food, 2, 5, 41, 45, 100, 108
food production, 108
force, 68, 76
freedom, 44
freshwater, 57
fruits, 73
fungal infection, 93, 100

G

garbage, 62
general knowledge, 30
geography, 63, 64, 65, 113

Germany, 88, 89, 93, 94, 95, 96, 98, 99, 100, 101, 102, 103, 105, 106, 107, 108, 110, 111, 112, 113
Gestalt, 100
GIS, 113
global warming, 57
goal of production, viii, 8
goal setting, 30, 102
governments, 67
grading, 72
gravity, 76
Greece, 111
green production, viii, 42
green products, viii
grids, 24
groundwater, 61
guidelines, 67, 80

H

happiness, 44
harvesting, 16, 41, 71
health, 2, 65, 85, 93
health information, 93
height, 76
higher education, 85
hiring, 33
history, 41, 63, 64, 65
horticultural crops, 2
House, 96, 103
human, vii, viii, 1, 3, 5, 9, 23, 24, 25, 26, 27, 29, 30, 31, 32, 33, 39, 41, 42, 43, 44, 45, 55, 56, 57, 63, 64, 66, 67, 68, 70, 76, 77, 79
human actions, viii, 55
human activity, 43
human development, 66
human health, 45, 79
Human Resource Management, 33, 88, 111, 112
human resources, 29, 39
human welfare, 44
humanitarian aid, 9
humidity, 65

I

identification, 67, 93
identity, 94
image, 13, 16, 37, 38, 100, 104
images, 81
impact assessment, 56, 57, 107
Impact Assessment, 92, 101, 107
implementation process, ix, 23, 25, 27, 39
imported products, 8
improvements, 18
impurities, 72
income, 5, 8, 26, 51, 52, 61, 68, 71, 72, 75, 79, 81, 82, 83, 84
independence, 30
India, 85, 86, 87, 94, 95, 98, 100, 103, 105, 106, 107, 108, 109, 110, 112, 113
individuals, 66, 67
industry, 1, 5, 38, 42, 46, 47, 111
information processing, 26
information technology, 32
infrastructure, 23, 43, 46, 51
injury, 25, 77
institutions, 68
intellect, 66
intelligence, 66
interference, 9
international relations, 65
interpersonal relations, 66
interpersonal relationships, 66
intervention, 70, 78
investments, 27
Iowa, 94
Iran, 41, 85, 98, 99, 100, 101, 104, 106, 108, 114
iron, 6
irrigation, 90, 105
Islam, 109
issues, viii, 9, 39, 59, 82

J

journalists, 38

L

labeling, 81
landscape, 96
landscapes, 43
languages, 14
laws, 67, 68
leadership, 30
learning, 25, 34, 64, 85
life cycle, 5, 6, 56, 95, 99, 100, 106
life quality, 45
lifetime, 77, 78
livestock, 2
living conditions, 42
logistics, 81
longevity, 78, 91
lower prices, 50
loyalty, 25
lying, 25

M

machinery, 2, 13, 29, 31, 36, 70, 77
Mackintosh, 25, 102
macroeconomics, 49
magnetism, 55
magnitude, 71
management, v, ix, 8, 21, 29, 30, 31, 32, 33, 34, 35, 36, 37, 38, 39, 42, 47, 59, 60, 85, 86, 87, 88, 89, 90, 91, 92, 93, 94, 95, 96, 97, 98, 99, 100, 101, 102, 103, 104, 105, 106, 107, 108, 109, 110, 111, 112, 113, 114
manufacturing, ix, 2, 7, 8, 18, 22, 31, 33, 66, 67, 71, 73, 75, 76, 82, 83
marketability, 31, 64, 69, 72, 79, 83, 84
marketing, 31, 34, 36, 64, 81
marketplace, 71
Maryland, 98, 102
mass, 7, 69, 72, 75, 76
materials, 2, 3, 6, 7, 12, 23, 26, 27, 29, 31, 35, 39, 45, 57, 58, 61, 62, 68, 71, 72, 80, 81
mathematics, 14, 66
matter, 73
measurement, 79

measurements, 105
meat, 81
media, 80
medical, 2, 7, 79
Mediterranean, 87
MEMS, 103
mental processes, 66
mentor, 30
metals, 26
methodology, 56
microeconomics, 49
microorganisms, 55
military, 65
mineral water, 81
Missouri, 100, 105, 112
Montana, 98
motivation, 9, 30, 66
multimedia, 24
music, 64

N

National Research Council, 69
natural gas, 26, 61
natural resources, 2, 38, 39, 41, 42, 43, 47, 55, 59, 62, 75, 99
natural science, 66
negative effects, 62
negotiation, 12
Netherlands, 86, 89, 90, 95, 97, 100, 103, 104, 105, 109, 114
neural network, 100
neural networks, 100
next generation, 43

O

OECD, 92
oil, 2, 77
operations, 7, 9, 31, 32, 46, 69, 72
opportunities, 30, 36
oppression, 25
optimization, 17, 18
ores, 61
organize, 8
organs, 103

ox, 87
oxidation, 57
ozone, 57
ozone layer, 57

P

peace, 9, 67
performance indicator, 100
personal development, 44
personal responsibility, 30
personality, 66
petroleum, 26, 61
pharmaceutical, 79
Philadelphia, 112
physical features, 65
physical phenomena, 55
physical structure, 23
plants, 2, 65
plastics, 26, 61
point of origin, 35
poison, 77
Poland, 87
policy, 21, 46, 47, 67, 68
political parties, 67
politics, 67
pollutants, 4, 39, 58, 59, 81
pollution, 9, 55, 58
population, 41
population growth, 41
portfolio, 33
portfolio management, 33
potato, 93, 104, 114
preservation, 67
president, 30
prevention, 37
principles, 42, 76, 98
problem-solving, 14, 16, 25
procurement, 31, 36
producers, viii, 12, 27, 39, 41, 42, 51, 64, 65, 66, 67, 68, 71, 73, 74, 78, 79, 83
product design, 12, 53
product life cycle, 5
product market, 75, 105
production costs, 9, 31, 39, 50, 52, 53

profit, 29, 31, 36, 83
programming, 29, 31, 39
project, 31, 34
prosperity, vii, 41
protection, 62
prototype, 11, 18
psychology, 63, 64
public sector, 67
punishment, 65, 68
purchasing power, viii

Q

quality assurance, 37, 81
quality control, 74

R

radiation, 55
random errors, 74
raw materials, 2, 6, 8, 35, 61, 97
reading, 41
reality, 41, 66
recommendations, 18, 37, 57
recovery, 39, 77
recruiting, 33
recycling, 5, 6, 39, 42, 55, 56, 58, 62, 81, 95
regulations, 68
reliability, 69, 71, 74
religion, 44, 64, 65, 66
renewable energy, 26, 38, 42, 61
renewable fuel, 26, 43
rent, 27
repair, 61
reprocessing, 61
requirements, ix, 11, 14, 16, 19, 21, 23, 24, 27, 29, 39, 65, 67, 73, 77
RES, 61
researchers, vii, 41
resilience, 30, 42, 66
resistance, 76
resource management, 33, 114
resources, 8, 32, 34, 39, 41, 42, 43, 45, 60, 61, 67, 76, 78, 79, 81, 97
response, 100

rights, 25, 66
risk, 31, 36, 37, 73
risk management, 37
risks, 32, 37
risk-taking, 31
root, 64
Royal Society, 96
rules, 2, 68

S

safety, 25, 33, 39, 55, 69, 70, 71, 73, 77
SAS, 14
scarce resources, 39
science, 34, 43, 65, 67, 80, 114
scientific method, 34
scope, 56
seafood, 46
securities, 36
security, 82
self-awareness, 25
self-improvement, 30
seller, 83
sellers, 4, 50, 83, 84
services, viii, 2, 4, 5, 11, 18, 32, 35, 36, 43, 49, 50, 69, 75, 79, 83
shape, 71, 72, 75, 100
shelf life, 69, 79, 81
shelter, 64
Singapore, 89, 105, 108, 114
social behavior, 64
social capital, 65
social order, 65
social problems, 9, 10
social responsibility, 26
social sciences, 2
society, vii, 24, 64, 65, 66, 67, 70, 90, 91, 112
sociology, 63, 65, 109
software, 4, 11, 13, 14, 17, 22, 24, 30, 57
Solomon I, 94
South Africa, 93, 111
Spain, 106
species, 43
specifications, 73

spending, 67
state, vii, 41, 43, 71, 73, 108
statistics, 14
statutes, 68
stimulus, 59
STM, 72, 86
stock, 43
storage, 69, 71, 79, 82, 84
strategic management, 31
stratification, 65
stress, 30
stress management, 30
structure, 2, 99
subjective experience, 66
supervisors, 30
supplier, 84
supply chain, 1, 31, 36
survival, 43
sustainability, v, vii, viii, ix, 41, 42, 43, 44, 45, 46, 47, 49, 53, 55, 56, 58, 59, 65, 69, 74, 80, 81, 82, 85, 86, 87, 88, 89, 90, 91, 92, 93, 94, 95, 96, 97, 98, 99, 100, 101, 102, 104, 106, 107, 108, 109, 110, 111, 112, 113
sustainable development, 45, 93
Sustainable Development, 44, 89, 90, 103, 111
sustainable environment, viii
sustainable production, vii, viii, ix, 1, 11, 12, 41, 46, 47, 49, 50, 53, 55, 56, 58, 62, 63, 64, 66, 68, 69, 70, 74, 75, 84
Switzerland, 85, 87, 88, 89, 91, 92, 97, 99, 101, 102, 103, 105, 107, 108

T

tactics, 37
target, 16
task performance, 102
taxes, 68
teams, 32, 37
techniques, 16, 24, 31, 37, 70
technologies, 5, 14, 24, 33, 35, 42, 70, 90, 114

technology, 1, 3, 5, 8, 23, 24, 27, 29, 31, 33, 34, 35, 39, 42, 43, 64, 69, 70, 80, 99, 100, 103, 112
technology transfer, 33
telecommunications, 24
temperature, 16
testing, 7, 18, 74, 107
theft, 25
thoughts, 9, 30, 34, 44, 64
tides, 60
tourism, 46
toxicity, 57
Toyota, 104
trade, 49, 73
training, 24, 31, 73
traits, 25
transport, 2, 80
transportation, 1, 4, 36, 42, 80, 82
Turkey, 87, 100

U

uniform, 72
unit cost, 107
United Kingdom, 85, 100, 107
United States, 91
universe, viii
universities, 14, 16, 30
urban, 46
USA, 94, 97, 109, 113

V

Valencia, 106
valuation, 106
variables, 16, 27, 50
varieties, 104
vegetables, 73, 81
vegetation, 55, 114
vibration, 59, 77
vision, 30

W

wages, 3
Washington, 89, 94, 103, 109
waste, 6, 32, 39, 61, 62, 95
waste management, 32
water, 4, 6, 24, 26, 38, 42, 55, 58, 77, 81, 90, 114
waterways, 82
wealth, 49
wear, 71, 73
web, 12
webpages, 80
weight reduction, 90
welfare, 8, 39, 42, 44, 45, 65
well-being, 43, 44, 45
wellness, 33
wood, 41
word processing, 14
workers, 26
workflow, 32
working conditions, 24
workstation, 7
World Bank, 111

Y

yield, 52, 53, 58, 60